Wireless Communication Standards

A Study of IEEE 802.11™, 802.15™, and 802.16™

Todor Cooklev

Published by
Standards Information Network
IEEE Press

Trademarks and Disclaimers

IEEE believes the information in this publication is accurate as of its publication date; such information is subject to change without notice. IEEE is not responsible for any inadvertent errors.

Library of Congress Cataloging-in-Publication Data

>   Cooklev, Todor, 1966-
>   Wireless communication standards : a study of IEEE 802.11, 802.15, and 802.16 / Todor Cooklev.
>   p. cm.
>   Includes bibliographical references and index.
>   ISBN 0-7381-4066-X
>   1. IEEE 802.11 (Standard) 2. wireless communication systems--Standards. I. Title.
>   TK5105.5668.C66 2004
>   621.382--dc22
>   2004054876

IEEE
3 Park Avenue, New York, NY 10016-5997, USA

Copyright © 2004 by IEEE
All rights reserved. Published August 2004. Printed in the United States of America.

No part of this publication may be reproduced in any form, in an electronic retrieval system, or otherwise, without the prior written permission of the publisher.

Yvette Ho Sang, Manager, Standards Publishing
Jennifer Longman, Managing Editor
Linda Sibilia, Cover Designer

IEEE Press/Standards Information Network publications are not consensus documents. Information contained in this and other works has been obtained from sources believed to be reliable, and reviewed by credible members of IEEE Technical Societies, Standards Committees, and/or Working Groups, and/or relevant technical organizations. Neither the IEEE nor its authors guarantee the accuracy or completeness of any information published herein, and neither the IEEE nor its authors shall be responsible for any errors, omissions, or damages arising out of the use of this information.

Likewise, while the author and publisher believe that the information and guidance given in this work serve as an enhancement to users, all parties must rely upon their own skill and judgement when making use of it. Neither the author nor the publisher assumes any liability to anyone for any loss or damage caused by any error or omission in the work, whether such error or omission is the result of negligence or any other cause. Any and all such liability is disclaimed.

This work is published with the understanding that the IEEE and its authors are supplying information through this publication, not attempting to render engineering or other professional services. If such services are required, the assistance of an appropriate professional should be sought. The IEEE is not responsible for the statements and opinions advanced in the publication.

Review Policy

The information contained in IEEE Press/Standards Information Network publications is reviewed and evaluated by peer reviewers of relevant IEEE Technical Societies, Standards Committees and/or Working Groups, and/or relevant technical organizations. The authors addressed all of the reviewers' comments to the satisfaction of both the IEEE Standards Information Network and those who served as peer reviewers for this document.

The quality of the presentation of information contained in this publication reflects not only the obvious efforts of the authors, but also the work of these peer reviewers. The IEEE Press acknowledges with appreciation their dedication and contribution of time and effort on behalf of the IEEE.

To order IEEE Press Publications, call 1-800-678-IEEE.

Print: ISBN 0-7381-4066-X   SP1134

See other IEEE standards and standards-related product listings at: http://standards.ieee.org/

Trademarks

3Com is a registered trademark of 3Com Corporation (www.3com.com/).

Bluetooth is a registered trademark of Bluetooth SIG, Inc.(www.bluetooth.org/).

Cisco is a registered trademark of Cisco Systems, Inc. (www.cisco.com/).

Ericsson is the trademark or registered trademark of Telefonaktiebolaget L M Ericsson, Sweden (www.ericsson.com/).

Fujitsu is a registered trademark of Fujitsu Limited in Japan and/or other countries (www.fujitsu.com/).

IBM is a registered trademark of International Business Machines Corporation in the United States, other countries, or both (www.ibm.com/).

IEEE and 802 are registered trademarks of the IEEE (www.ieee.org/).

IEEE Standards designations are trademarks of the IEEE (www.ieee.org/).

Intel is a registered trademark of Intel Corporation (www.intel.com/).

Kodak is a trademark of Eastman Kodak Company (www.kodak.com/).

Lucent is a registered trademark of Lucent Technologies Incorporated (www.lucent.com/).

Microsoft is a registered trademark of Microsoft Corporation (www.microsoft.com/).

Motorola is a registered trademark of Motorola, Inc. (www.motorola.com/).

Nokia is a registered trademark of Nokia Corporation (www.nokia.com/).

SOMA Networks is a trademark of SOMA Networks, Inc. (www.somanetworks.com/).

Staccato Communications, Inc. is a registered trademark of Staccato Communications, Inc. (www.staccatocommunications.com/).

Texas Instruments is a trademark of Texas Instruments Incorporated (www.ti.com/).

Time Domain is a registered trademark of Time Domain (www.time-domain.com/).

Toshiba is a registered trademark of Toshiba Corporation (www.toshiba.com/).

Wi-Fi is a registered trademark of the Wi-Fi Alliance (www.wi-fi.org/).

WiMAX is a trademark of the WiMAX Forum (www.wimaxforum.org/).

XtremeSpectrum is a registered trademark of Motorola, Inc. (www.motorola.com/)

ZibBee Alliance is a trademark of Philips Corporation (www.zigbee.org/).

Dedication

To my wife Didi and my son Steven

Acknowledgment

A project like this book cannot be accomplished without the help of others, whom I would like to publicly acknowledge here. First, it has been personally and professionally rewarding to work with the many outstanding professionals within IEEE 802®. Second, I would like to acknowledge those I have worked closely with in industry over the years. I have gained substantially from interaction with former colleagues and friends Marcos Tzannes and Colin Lanzl. I want to also acknowledge former colleagues and friends Kurt Dobson, Dirk Ostermiller, Sy Prestwich, and Dr. Kevin Smart. I want to thank Dr. Gheorghe Berbecel for his friendship and the many stimulating discussions. I am also thankful to Victor Mozarowski for his friendship and discussions. I want to also acknowledge Dr. M. Sablatash for the interaction I have had with him over the years.

I want to thank also all of my friends. And last, but not least, I am thankful to my family, since they have made what I have achieved possible.

I am also grateful to Jennifer Longman from the IEEE Standards Information Network/IEEE Press for her work and above all for her patience during this two-year project.

I acknowledge Stefano Iachella for proof-reading the text.

I want to acknowledge specifically the anonymous reviewers from the IEEE for their constructive criticism of the completely inadequate first draft. They should be credited with all the improvements.

I am particularly indebted to the following people: Dr. Steve Halford, Globespan Virata, Palm Bay, FL; Dr. John Kowalski, Sharp Laboratories of America, Camas, WA; Dr. Martin Haenggi, University of Notre Dame, IN; and Dr. Sean Coffey, Texas Instruments, Santa Rosa, CA, for reviewing the manuscript and for their detailed comments.

Author

Todor Cooklev has been with San Francisco State University since 2002 where he teaches courses and conducts research in the area of wireless communications. During 1999–2002 he was with Aware, Inc. of Bedford, MA and Lafayette, CA. Prior to joining Aware he was with US Robotics, later 3Com Corporation, where he received the 3Com Inventor Award. In addition to wireless communications, he has worked in the area of voice band modems, DSL technology, and digital signal processing. He also has experience in consulting for government organizations and private technology and venture capital companies.

Dr. Cooklev has given a number of short courses, seminars, and invited talks and has also made contributions to the IEEE 802.11, 802.15, and 802.16 families of standards. He is a member of the IEEE Standards Association and has served the IEEE in a number of volunteer capacities. He is also on the Advisory Board of the IEEE International Conference on Telecommunications.

He received the Ph.D. degree in electrical engineering from Tokyo Institute of Technology, Tokyo, Japan, in 1995, and also received a NATO Science Fellowship Award in that same year. He is the inventor on several patents in the United States and author of a number of publications in the areas of communication systems, protocols, and signal processing.

Table of Contents

Preface xi

Acronyms and Abbreviations xiii

Chapter 1 Introduction 1
  Overview 1
  Government regulations 5
    Recent regulatory changes 10
    Future FCC directions. 12
    International Regulations 13
  Standardization bodies 17
  Wireless channels 24
  Introduction to cryptographic algorithms 32
  Design of a wireless communication standard 35
    MAC requirements 36
    PHY requirements 40
    Sublayers of PHY and MAC 41

Chapter 2 The IEEE Standard for WLAN: IEEE 802.11 45
  Overview and architecture 45
  IEEE 802.11 security 49
    Authentication 50
    Privacy 58
  Interaccess Point Protocol (IAPP) 66
  Medium-access mechanism and real-time traffic over
  IEEE 802.11 70
    Enhancements in IEEE 802.11e 81
  Power-saving mechanism 94
  IEEE 802.11 physical layers 96
    DSSS PHY 97
    IEEE 802.11b 98
    IEEE 802.11a 103
    IEEE 802.11g 113

WLAN installation 117
IEEE 802.11 technology and business trends 125

Chapter 3 Standards for Wireless Personal Area Networking (WPAN) . 133
Introduction 133
Comparing WPAN and WLAN 133
IEEE 802.15.1 135
    IEEE 802.15.1 Physical layer................................ 137
    Baseband 139
    LMP and L2CAP 162
Coexistence among wireless standards 164
    Collaborative methods 168
    Noncollaborative methods 170
    Channel classification...................................... 174
    PHY and MAC models 176
High-Rate WPAN 179
    MAC layer 181
    Physical layer for the 2.4 GHz ISM Band 195
Low-Rate WPAN 201
    Overview and architecture 201
    PHY Layer 204
    IEEE 802.15.4 MAC 207
    Coexistence issues involving IEEE 802.15.4.................... 216
WPAN technology and business trends 218

Chapter 4 Air Interface for Fixed Broadband Wireless Access Systems . 225
MAC convergence sublayer 231
MAC common part sublayer 233
    Network entry and initialization............................. 238
    Channel access and QoS.................................... 241
MAC security sublayer 248
    Authentication... 250
    Data encryption with DES 255
MAC enhancements for 2–11 GHz operation 256
    MAC enhancements for mesh systems 257
    Advanced antenna systems (AAS)............................ 261

Automatic repeat request (ARQ) 265
DFS for license-exempt operation 267
MAC enhancements for OFDM and OFDMA PHYs. 268
IEEE 802.16 physical layers 269
Physical layer for 10–66 GHz. 270
Physical layers for 2–11 GHz operation. 286
Coexistence 317
BWA business and technology trends 331

Chapter 5 Concluding Remarks 335

Bibliography 337

Glossary 351

Index 357

Preface

The purpose of this book is to discuss the design of the standards for wireless data communications. In particular, the book strives to answer the questions: How do these standards work? Why do they work the way they do? and How were they developed? The field of computer networking emerged with the advent of computers. The first mass-market data communication standard was Ethernet, standardized as IEEE 802.3™. Ethernet is a technology allowing computers to communicate over relatively short distances, such as offices, and campuses, and is the dominant wireline local-area networking technology (LAN). Ethernet requires special wiring such as CAT5 cables that are not available in homes and some buildings. The success and limitations of Ethernet, as well as the proliferation of laptops and mobile computing devices created significant market demand for wireless computer networks. The standards for broadband wireless data communication were developed in response to these market needs. These standards aim to provide integrated, packet-oriented transmission of data, voice, images, and video. These wireless technologies will have profound impact. In the near term, these wireless standards will create and expand markets for products and services. They will facilitate trade and commerce. In the long term, these wireless standards will lead to increased economic productivity and competitiveness. At present, the market for wireless data communications is one of the fastest growing communications markets, even as many other communication markets are in decline. As a result, even technology and service provider companies that have traditionally offered products and services for wired communication technology are beginning to offer products involving wireless communication technology. This confirms the shift from wired to wireless communication technology that is presently occurring. For many low-tech companies the use of wireless technology is new and results in substantially increased productivity. It is expected that the proliferation of laptops and, especially, of other mobile computing devices such as PDAs and home networking will continue to contribute to the growth in wireless data communications over the foreseeable future.

Almost all products and services currently being introduced are based on the standards developed by the IEEE 802 LAN/MAN Standards Committee. As a result many engineers and computer scientists, students, marketing specialists, and managers must educate themselves about these standards. This book is designed to help practicing professionals learn about the standards. Graduate students doing research work in the area of wireless data communication technology will also benefit from using this book.

I have attempted to make the text accessible to a wide audience. The concepts are explained based on intuition and insights. Background in signal and system theory, modulation and coding, while helpful, is not particularly required. The book is suitable for novice readers who want to learn more about the subject, as well as for advanced readers who are working in wireless communications. The organization is as follows. In the introductory Chapter 1 the main requirements for the wireless standards (802.11™, 802.15™, and 802.16™) and the main technical difficulties that the wireless environment creates are discussed. The important issues of government regulations and the standardization process are also explained. Chapter 2 is devoted to the IEEE 802.11 standards for wireless local area networks, which historically were the first wireless data communication technology to be developed. The practical problem how to deploy wireless local area networks is also discussed. The IEEE 802.15 family of standards for wireless personal area networking is the subject of Chapter 3, and the IEEE 802.16 standard for broadband wireless access is the subject of Chapter 4. In general there are two ways to view any communications system: architectural and functional. The architectural approach emphasizes the logical divisions of the system and how they work together. The functional approach emphasizes the actual components. In each chapter, first the architecture is presented, followed by a description of the actual components.

It should be noted that the technical subject of this book is rapidly evolving. I have made an effort within the text to reflect not only the state of the wireless standards at publication time (2004), but to provide guidance as to what developments might happen in the future. It should be noted that the views expressed here represent solely my personal opinion. They do not necessarily coincide with the positions of either the IEEE 802 Committee, or the individual Working Groups or Task Groups.

Acronyms and Abbreviations

The following acronyms and abbreviations are used in this book:

2G	second generation
3-DES	triple data encryption standard
3G	third generation
AA	authorization agent
AAA	authentication, authorization, and accounting
AAD	additional authentication data
AAS	advanced antenna systems
AC	access category
ACK	acknowledgement
ACL	asynchronous connection-less or access control list
ACO	authenticated ciphering offset
ADSL	asymmetric digital subscriber line
AES	advanced encryption standard
AFH	adaptive frequency hopping
AGC	automatic gain control
AIFS	arbitration inter-frame space
AIFSN	arbitration inter-frame space number
AK	authorization key
AP	access point
APME	access point management entity
ARIB	Association of Radio Industries and Businesses
ARQ	automatic repeat request
AS	authentication server
ASTM	American Society for Testing and Materials
ATIM	ad-hoc traffic indication message
ATM	asynchronous transfer mode
AWMA	alternating wireless medium access
AWGN	additive white Gaussian noise

BB	baseband
BCC	binary convolutional code
BE	best effort
BER	bit error rate
BIFS	backoff inter-frame space
BPSK	binary phase shift keying
BTC	Block turbo code
BS	base station
BSS	basic service set
BSSID	basic service set identifier
BT	3 db bandwidth-bit duration product for Gaussian frequency shift keying
BTC	block turbo code
BW	bandwidth
BWA	broadband wireless access
CA	collision avoidance
CAT5	category 5
CAP	contention access period
CAZAC	constant amplitude zero autocorrelation
CBC	cipher block chaining
CBC-MAC	cipher block chaining-message authentication code
CBR	constant bit rate
CC	controlled contention
CCA	clear channel assessment
CCI	controlled-contention interval
CCITT	(French) International Telegraph and Telephone Consultative Committee (now ITU)
CCK	complementary code keying
CCM	counter + cipher block chaining-medium access control
CCMP	counter + cipher block chaining-medium access control protocol
CCS	common channel signaling
CD	collision detection or compact disc
CDMA	code-division multiple access

Acronyms and Abbreviations

CEI	channel estimation interval
CEPT	European Conference of Postal and Telecommunications Administrations
CF	coordination function
CFB	contention-free burst
CFP	contention-free period
CFR	Code of Federal Regulations
CINR	carrier-to-interference-plus-noise-ratio
COF	ciphering offset number
CP	contention period or cyclic prefix
CRC	cyclic redundancy code
CS	convergence sublayer
CSMA/CA	carrier sense multiple access with collision avoidance
CTA	channel time allocation
CTC	convolutional turbo code
CTR	counter
CTS	clear to send
CVSD	continuously variable slope delta modulation
CW	contention window
DAMA	demand assigned multiple access
DBPSK	differential binary phase shift keying
DC	direct current
DCD	downlink channel descriptor
DCS	dynamic channel selection
DES	digital encryption standard
DEV	device
DFS	dynamic frequency selection
DFT	discrete Fourier transform
DHCP	dynamic host control protocol
DIAC	dedicated inquiry access code
DIFS	distributed inter-frame space
DL	downlink
DLP	direct link protocol

Acronyms and Abbreviations

DQPSK	differential quadrature phase shift keying
DS	direct sequence
DSL	digital subscriber line
DSRC	dedicated short-range communication
DSSS	direct sequence spread-spectrum
DVD	digital video disc
EAP	extensible authentication protocol
EAPOL	extensible authentication protocol over local area network
EC	elliptic curve
ECB	electronic codebook mode
ECC	elliptic curve cryptography
ECDSA	elliptic curve digital signature algorithm
ECMQV	elliptic curve Menezes-Qu-Vanstone
EDCF	enhanced distributed coordination function
EIA	Electronic Industry Association
EIFS	extended inter-frame space
EIRP	effective isotropic radiated power
EN	European norm
ERC	European Radio Commission
ESS	extended service set
ETSI	European Telecommunication Standardization Institute
FCC	Federal Communications Commission
FCH	frame control header
FCS	frame check sequence
FDD	frequency-division duplexing
FEC	forward error correction
FFT	fast Fourier transform
FH	frequency hopping
FIPS	Federal Information Processing Standard
FTP	file transfer protocol
GF	Galois field
GFSK	Gaussian frequency shift keying
GIAC	general inquiry access code

GTS	guaranteed time slot
HEC	header error check
HC	hybrid coordinator
HCCA	hybrid coordination function controlled channel access
HCF	hybrid coordination function
HCI	host control interface
HDR	high data rate
H-FDD	half duplex frequency division
HIPERLAN	high performance radio local area network
HR	hop range
HUMAN	high-speed unlicensed metropolitan area network
HVAC	heating, ventilating, and air conditioning
I	inphase
IA	interference area
IAPP	inter-access point protocol
IBSS	independent basic service set
IDFT	inverse discrete Fourier transform
IEC	International Electrotechnical Commission
IETF	Internet Engineering Task Force
IEEE	Institute of Electrical and Electronic Engineers, Inc.
IF	intermediate frequency
IFFT	inverse fast Fourier transform
IFS	inter-frame space
IP	internet protocol
ISI	intersymbol interference
ISM	industrial, scientific, medical
ITU	International Telecommunications Union
JTC	joint technical committee
KEK	key encryption key
L2CAP	logical link control and adaptation protocol
LAN	local area network
LC	link controller
LFSR	linear feedback shift register

LLC	logical link control
LMP	link manager protocol
LMSC	Local and Metropolitan Area Network Standards Committee
LOS	line of sight
MAC	medium access control
MAN	metropolitan area network
MIB	management information base
MIC	message integrity code
MIFS	minimum inter-frame space
MIMO	multiple-input multiple-output
MLME	medium access control layer management entity
MMDS	multichannel multipoint distribution service
MP-MP	multipoint-to-multipoint
MSDU	medium access control service data unit
MWS	multimedia wireless systems
NAV	network allocation vector
NIST	National Institute of Standards and Technology
NLOS	non-line of sight
NOI	Notice of Inquiry
NPRM	Notice of Proposed Rulemaking
nrtPS	non-real-time polling service
NTIA	National Telecommunications and Information Administration
OCB	offset codebook mode
OFDM	orthogonal frequency division multiplexing
OFDMA	orthogonal frequency-division multiple access
OQPSK	offset quadrature phase shift keying
OSI	open system interconnection
PAN	personal area network
PBCC	packet binary convolutional coding
PBX	private branch exchange
PC	personal computer or point coordinator
PCF	point coordination function

Acronyms and Abbreviations

PCM		pulse code modulation
PCS		personal communication service
PDU		protocol data unit
PER		packet error rate
PF		persistence factor
PHS		payload header suppression
PHY		physical layer
PIFS		point coordination function inter-frame space
PIN		personal identification number
PKM		privacy key management
PL		path loss
PLL		phase locked loop
PLME		physical layer management entity
PMD		physical medium dependent
PMK		pairwise master key
PMP		point-to-multipoint
PN		packet number or pseudo-noise
PNC		piconet coordinator
PPM		pulse position modulation
PS		power save
PSD		power spectral density
PTA		packet traffic arbitration
PSPS		piconet synchronized power save
PTA		packet traffic arbitration
PTK		pairwise transient key
Q		quadrature
QAM		quadrature amplitude modulation
QLRC		quality long retry count
QoS		quality of service
QPSK		quadrature phase shift keying
QSRC		quality short retry count
RADIUS		remote authentication dial-in user service
RF		radio frequency

RFC	request for comments
RIFS	retransmission inter-frame space
RLAN	radio local area network
RPE	radiation pattern envelopes
RRM	radio resource management
RS	Reed–Solomon or repeater station
RSA	Rivest, Shamir, and Adleman
RSSI	received signal strength indicator
rtPS	real-time polling service
RTS	request to send
RTTT	road transport and traffic telemetric
RX	receiver
SA	security association
SAP	service access point
SB	stuff bits
SC	single-carrier
SCO	synchronous connection-oriented
SDL	specification description language
SDP	services discovery protocol
SDR	software defined radio
SDU	service data unit
SIFS	Short inter-frame space
SIG	Special Interest Group
SINR	signal-to-interference-plus-noise-ratio
SLP	service location protocol
SNMP	simple network management protocol
SNR	signal-to-noise ratio
SME	station management entity
SP	service period
SPS	synchronous power save
STC	space-time coding
SS	Subscriber station
SSID	SS identifier

Acronyms and Abbreviations

TBTT	targeted beacon transmission time
TC	traffic category
TCM	Trellis-code modulation
TCP	transmission control protocol
TCS	telephony control signaling
TDD	time-division duplexing
TDMA	time-division multiple access
TE	terminal equipment
TEK	traffic encryption key
TFTP	trivial file transfer protocol
TIM	traffic indication map
TKIP	temporary key integrity protocol
TPC	turbo product code or transmitter power control
TS	traffic stream
TSC	temporary key integrity protocol sequence number
TSPEC	traffic specification
TBTT	targeted beacon transmission time
TV	television
TX	transmitter
TXOP	transmission opportunity
UCD	uplink channel descriptor
UDP	user datagram protocol
UGS	unsolicited grant service
UL	uplink
UNII	unlicensed national information infrastructure
USB	universal serial bus
UW	unique word
UWB	ultra-wide band
VOIP	Voice over IP
WAN	wide-area network
WECA	Wireless Ethernet Compatibility Alliance (now Wi-Fi)
WEP	Wireless equivalent privacy
Wi-Fi	wireless fidelity

WLAN	wireless local area network
WPA	wireless protected access
WPAN	wireless personal area network
WRC	World Radio Congress
XOR	Exclusive "or"
XPD	cross-polar discrimination

Chapter 1 Introduction

People often take the view that standardization is the enemy of creativity. But I think that standards help make creativity possible—by allowing for the establishment of an infrastructure, which then leads to enormous entrepreneurialism, creativity, and competitiveness.

—Vint Cerf, *TCP/IP co-developer and Internet pioneer*

OVERVIEW

Wireless data communications technology is incessantly progressing from research to standardization to implementation. This book proposes that this process (i.e., the progress of communication technology) is governed by the following four main principles. First, Shannon's law:

$$C = W\log_2\left(1 + \frac{P}{N_0 W}\right) \qquad \text{Eq. 1-1}$$

where C is the capacity of a communications system in bits per second, P is the signal power in Watts, W is the signal bandwidth in Hertz, and N_0 is the one-sided noise power spectral density in Watts/Hz. Shannon's law says that if there is one transmitter and one receiver, the capacity of a communication channel depends linearly on the available bandwidth and logarithmically on the signal-to-noise ratio. In other words, it is more difficult to increase the capacity of a channel by increasing the signal-to-noise ratio than it is by increasing the available bandwidth.

The second fundamental principle is Moore's law, which states that the level of integration of integrated circuits will double every 18 months.

The third principle is that the value of a network is proportional to the square of the speed of connection.[1]

---

[1] This principle is sometimes attributed to R. Metcalfe, inventor of Ethernet and founder of the data networking company 3Com® Corporation.

Chapter 1: Introduction

The fourth main principle is that the value of a network is proportional to the square of the number of devices that can be connected. With respect to a particular device, the fourth principle says that the value of a device is proportional to the square of the number of devices with which it can communicate.

The place of Shannon's law among this set of main principles does not need to be justified. Communications technology, the wireless standards in this book included, has been trying to get closer to Shannon's capacity since Shannon's work in 1948. Moore's law is very well known in the area of integrated circuit design. Why is it also a communications law? It is a communications law because it provides the foundation to implement increasingly complex and powerful signal processing circuits. In this way, Moore's law effectively allows increasingly complex and powerful wireless devices to be implemented. Companies and businesses, of course, want constantly to increase the value of their networks and devices. While the first two principles address the question of what can be accomplished, the third and fourth principles address the question of how this increase in value can be accomplished—by increasing the rate of communication and by increasing the number of attached devices. The development of the IEEE 802® wireless data communication standards is a good manifestation of these fundamental laws of communications. The wireless standards provide the technology to interconnect a large number of devices at increasing data rates.

Before describing the standards for wireless data communications, the main characteristics of the ideal wireless technology should be identified. These characteristics include the following:

a) Ability to transport voice, audio, and video, in addition to computer data

b) Allowing devices with differing price, power consumption, and data rates to operate

c) Efficiently and dynamically allocating spectrum among the various networked devices

These requirements deserve to be discussed in greater detail.

Historically, much like the situation in wired networks, the development of wireless technology proceeded along two main paths—voice and data networks. Voice-oriented networks were developed first, followed by data-oriented networks. The main reason is business—the market for voice-oriented products and services developed earlier. However, at present there are two notable long-term market trends. First, the market for data-oriented products and services is growing considerably faster than the market for voice-oriented products and services. Second, voice-oriented and data-oriented products and services are merging. As a result, voice-oriented networks have increased data capabilities, and modern data-oriented networks can efficiently transport voice. One consequence of these market trends is the increased competition between vendors of voice-oriented products and vendors of data-oriented products. Therefore, the ideal wireless technology must efficiently transport voice, audio, and video in addition to data. Voice, audio, and video have different requirements than data. For data, the most important parameter is throughput, but the delay (or latency) is not important. For voice and video, in addition to a minimum throughput, the signal delay is very important. Substantial progress has been made recently to satisfy this requirement. All wireless standards developed by the IEEE 802 Working Groups strive to transport voice, audio, and video, in addition to data.

The second and third requirements (item b and item c, previously) towards the ideal wireless technology are much more difficult to satisfy, despite many claims to the contrary. There even does not seem to be universal agreement about the meaning of the second and third requirements.

Almost every wireless technology supports multiple data rates. For example, IEEE Std 802.11b™ support data rates of 1, 2, 5.5, and 11 Mb/s, IEEE Std 802.11a™ supports data rates between 6 and 54 Mb/s, etc. However, IEEE Std 802.11a devices that can communicate only at 6 Mb/s would cost the same as devices that can communicate at 54 Mb/s. At present, wireless technologies provide data rate scalability, but they provide no cost scalability at the same time. It becomes apparent that there is no technology for wireless communication that can satisfy the second requirement.

The third requirement—frequency agility—is necessary because the electromagnetic spectrum is limited and shared with other wireless devices. As the number of different technologies increased, frequency agility became important. Recent standards for wireless data communication have some form of frequency agility. The protocols have features to detect whether a frequency band is used by other devices, and if so, to coordinate the relocation to another frequency band. Still, standards lack complete flexibility. Frequency agility means truly dynamic spectrum allocation. Using the simple IEEE 802.11a example, a 6 Mb/s device will take as much bandwidth to communicate as a 54 Mb/s device.

It is obvious that existing technologies lack many of the characteristics of the ideal wireless technology. If a wireless technology had true scalability and frequency agility, then it would encompass all applications over short and long distances. The fact that such a technology does not exist has led to three main different wireless technologies—wireless local area networks (WLAN), wireless personal area networks (WPAN), and broadband wireless access (BWA) devices, also called wireless metropolitan area networks (WMAN).

The differences among WLAN, WPAN, and WMAN arise from the different design goals. WPAN devices are intended for very short-range communications—typically up to 10 m. In addition, they are used in small battery-powered devices and therefore must consume very little power. WLAN devices are intended for operation of up to 100 m and have intermediate power consumption requirements. Data rate for most WPAN devices is not as important as it is for WLAN devices, but battery life and cost are. As a result, WPAN devices achieve lower data rates and lower cost than WLAN devices. Some WPAN technologies being developed now aim for a different trade-off—very high data rates, but at very short distances. This trade-off is desirable to enable true "cable replacement." Both WPAN and WLAN devices work in unlicensed bands. WMAN devices are intended primarily to work over much larger distances—up to several kilometers. In addition, WMANs can work not only over unlicensed bands, but also over frequency bands that are licensed. WMAN devices are not intended to be used in battery-powered devices for extended periods of time.

The remainder of this chapter discusses government regulations relevant to the wireless data communication standards, followed by an overview of the standardization process. Because efficient design of wireless devices requires understanding of radio propagation, the characteristics of wireless channels are summarized in "Wireless channels" on page 24. An overview of the main technical characteristics of wireless standards is presented in "Design of a wireless communication standard" on page 35. The chapter concludes with an introduction to cryptographic systems which are used in current wireless communication standards. These standards are presented in subsequent chapters.

Chapter 1: Introduction

GOVERNMENT REGULATIONS

This section discusses why government regulations are necessary, what the relevant government regulations are in the United States and other major markets around the world, and how and in what direction these regulations are changing.

Arguably, spectrum is a precious resource in wireless communications. This situation is drastically different from wired communications where one can always run an extra wire, and spectrum is therefore essentially unlimited. The development of large markets for wireless products and services always depends, above all, on the availability of spectrum in appropriate frequency bands. The total radio spectrum comprises the region between 3 kHz and 300 GHz. Spectrum is of fundamental importance because from one frequency band to another, both radio propagation and the cost to build devices are different. In addition, antenna size depends on the frequency band of operation. In general, different frequency bands are appropriate for different systems. Furthermore, wireless devices are potentially subject to interference from other wireless devices that operate in the same frequency band. To maintain order, spectrum has been regulated by governmental organizations.

Another reason governments regulate the area of communications is the high cost of deploying equipment to provide nationwide coverage. In the U.S., the governmental organization that regulates spectrum for use by the Federal government is the National Telecommunications and Information Administration (NTIA), and the organization that regulates spectrum for commercial use is the Federal Communications Commission (FCC) [B55]. While the FCC rules officially govern operation only in the U.S., they are monitored by regulatory bodies all over the world, and even followed in varying degrees by many other nations in the Americas and the Pacific Rim.

 In the past, communications services in all countries were provided by monopolies. In the last 20 years there has been a move away from monopoly in the United States, followed by Canada, and to a lesser extent in other parts of the world. In the United States, the Telecommunications Act of 1996 played an especially big role in opening the communication markets to competition.

Chapter 1: Introduction

In the area of wireless data communications, the FCC has the following two main responsibilities:

a) Allocates spectrum for wireless data communication devices and establishes regulations, which the devices operating in the allocated spectrum must satisfy.

b) Certifies that the equipment on the U.S. market satisfies the regulations.

The demand for spectrum is significant, especially for certain frequency bands that have attractive propagation characteristics and allow inexpensive devices to be manufactured. Spectrum has been allocated for use by the military, law enforcement, and other government agencies, radio amateurs, etc. Similar to other regulatory bodies around the world, the FCC can respond to the demand for spectrum primarily by steady migration toward higher frequency bands, and—whenever possible and economically justified—by restructuring the frequency band allocations. Higher frequency bands are wider and therefore can accommodate more services. Restructuring of the frequency band allocations is seldom used because it is associated with significant cost.

There are two main types of frequency bands as per FCC regulations: licensed and unlicensed. Only the license holders are allowed to operate in licensed bands. Since 1994, these licenses have been awarded by auctions. The goal of the auction is to award the license to those who will use it most effectively and not to maximize revenue for the government. Nevertheless, economic efficiency does translate into higher revenue, and licenses are awarded to the highest bidder. No license is required for operation in the unlicensed bands. In the United States the FCC regulations are under the 47th part of the Code of Federal Regulations (CFR).

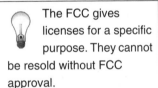
The FCC gives licenses for a specific purpose. They cannot be resold without FCC approval.

There are several specific FCC rules of high interest to designers of IEEE 802 class systems. These sections are 15.35, 15.205, 15.209, 15.247, and 15.249.

FCC 15.35 gives the requirements for detector and averaging functions for certification measurements. FCC 15.205 documents the *restricted bands* where only spurious emissions are allowed, and where those must meet the general levels of 15.209. Above 1,000 MHz, averaging according to 15.35 may be used.

Chapter 1: Introduction

FCC 15.209 restricts the RF energy that electronic equipment may parasitically emit. The specific level of emissions is 200 µV/m at 3 m test range below 960 MHz, and 500 µV/m above. These field strengths are approximately equivalent to –49.2 and –41.2 dBm ERP, respectively.

 This rule applies to the Industrial, Scientific, and Medical (ISM) bands, which are 902–928 MHz, 2400–2483.5 MHz, and 5725–5875 MHz.

FCC 15.247 is the primary category for U.S. operations of unlicensed equipment [B39]. This service category provides the potential for the highest performance of all the unlicensed service categories, allowing freedom from licensing, transmit powers up to 1 watt, and no limitations of application or transmit duty cycle. These requirements allow for carrier power up to 1 watt for direct sequence spread-spectrum systems in both 902–928 MHz and 2400–2483.5 MHz, if direct sequence processing gain is a minimum of 10 dB. However, effective May 30, 2002, the FCC eliminated the requirement for direct sequence spread-spectrum and replaced it by "digital modulation." Digital modulation was defined as the point when the combination of data rate, coding, and modulation method has a 6 dB bandwidth greater than 500 kHz and a maximum transmitted spectral density of less than +8 dBm/3 KHz. However, either direct sequence or frequency hopping spread spectrum may still be used to meet the requirements under 15.247 [B40].

In the same ISM bands of 902–928 MHz, 2400–2483.5 MHz, and 5725–5875 MHz, FCC 15.249 provides for narrowband (non–spread spectrum) operation of up to approximately 1 mW effective radiated power (if wideband, then 1 mW per 100 kHz below 1000 MHz and 1 mW/MHz above 1000MHz). Operation is allowed up to 50 mV rms of electric field strength from a transmitter at a 3-m test range. This is equivalent to 0.75 mW effective radiated power.

Other rules contained in CFR 47 provide allocations in the 260–470 MHz range (FCC 15.231), but because they are generally restricted on power, transmit duty cycle, and application, this band is primarily used for control and security applications (such as keyless entry). It is not of interest for high-speed wireless data communications.

The ISM bands are of particular interest to wireless data communications. These ISM bands were allocated to foster the development of new devices and service offerings that will stimulate economic development and the growth of new industries. The ISM regulations have indeed been extremely successful and many ISM devices are now available on the market. Because the spectrum is unlicensed, many more devices are yet to come. There are several reasons why these frequency bands are very attractive for wireless data communications. At these frequencies, a transmitter with power of 1 W or less with omnidirectional antenna can provide coverage up to 300 m indoors and up to a half a mile outdoors. At such frequencies, the antenna size is about one inch. At the ISM frequencies, today's semiconductor technology allows wireless devices to be built with reasonable power consumption, size, and cost. As the cost to build wireless transceivers at higher frequencies decreases, migration to higher frequencies and wider bands can be expected in the future.

IEEE 802 equipment would most often be certified under the FCC 15.247 service category [B39], [B40].

Devices that operate in the ISM bands must share the spectrum not only with other wireless data communication devices, but with other unlicensed devices. For example, the 902–928 MHz band has been used for cordless phones, and an increasing number of cordless phones work in the 2400–2483.5 MHz band. All these devices must share these frequency bands not only among themselves, but also with other government wireless systems. Many of these government systems are radiolocation systems, which transmit at high power levels. The intention of the rules is to prevent interference from the unlicensed devices to licensed systems, but not the opposite. Licensed systems are considered primary users and can cause interference to unlicensed devices. It is the responsibility of the vendor to design the unlicensed system so that it can operate in the presence of interference. Every company is allowed to develop and sell products operating in the unlicensed bands, provided that these products satisfy the FCC Part 15.247 regulations. Unlicensed devices can use frequency-hopping spread-spectrum (FHSS), direct-sequence spread-spectrum (DSSS), and novel digital modulation technologies. Spread-spectrum systems have less capability to produce harmful interference. Spread-spectrum systems work by spreading the digital signal by a pseudo-random code. The spreading process reduces the power spectral density of the transmitted signal.

Direct-sequence spread-spectrum devices work by first using conventional modulation and then multiplying the signal with a chipping sequence. Generally speaking spreading rates of at least $N = 10$ chips per symbol are required to achieve $10\log_{10} N = 10$ dB processing gain. To allow digital modulation schemes to be used, such as orthogonal frequency division multiplexing (OFDM), recent regulations eliminated the processing gain requirement and, instead, specified only the maximum power density [B40].

> The regulations before 2002 required direct-sequence spread-spectrum systems to have a processing gain of at least 10 dB. The 10 dB minimum was specified to ensure that the system is in fact spread spectrum.

Frequency-hopping systems work by changing or hopping the center frequency in a pseudorandom fashion. Therefore, frequency-hopping systems require a wider frequency band in which to hop. In the 902–928 MHz band, FH systems are allowed output power up to 1 W if they hop over 50 hopping channels of width 500 KHz, and 0.25 W if they hop over more than 24 but less than 50 channels. Frequency-hopping systems in the 2.4 GHz band are allowed output power of 30 dBm (or equivalently 1 W) if they hop over 75 channels 1MHz wide, and 0.125 W if they hop over more than 14 but less than 75 channels. In the 5.725–5.875 GHz band, FH systems are allowed output power up to 1 W. Note that the power in dBm is related to the power in megaWatts by

$$P_{dBm} = 10\log_{10} P_{mW}$$

and to obtain the power density, the signal power is divided by the occupied bandwidth. Allowing hopping to occur on less than the maximum number of channels is another recent modification of the rules. The new rules allow intelligent frequency-hopping systems to be built, which, by hopping on fewer channels, avoid other wireless systems operating in the same frequency band.

Until recently, only spread-spectrum systems were allowed, and digital modulation systems were not allowed in the ISM bands. The reason that the regulations in Part 15.247 were recently revised is that technological advances in digital modulation made possible systems that can cause less interference, much like spread-spectrum systems. At present in all three ISM bands digital modulation systems are allowed with output power levels of up to 1 W.

The power levels quoted previously are specified at the device-to-antenna connector and are valid for devices operating with antennas having gains of up to 6 dBi. If the antenna gain exceeds 6 dBi, the output power must be reduced. In the 902–928 MHz band, the output power must be reduced by 1 dB for every 1 dB of antenna gain above 6 dBi, and in the 2400–2483.5 MHz band, the output power must be reduced by 1 dB for every 3 dB of antenna gain above 6 dBi. For point-to-point systems in the 5.725–5.875 GHz band, highly directional antennas can be used with gains above 6 dBi without reduction of output power.

The ISM bands in the 900, 2400, and 5700 MHz regions are narrow. To enable high data rate wireless devices and services, in 1997 the FCC allocated a 300 MHz–wide block of frequencies for unlicensed use [B41], [B42]. These bands became known as unlicensed national information infrastructure (U-NII) bands and are from 5.15–5.25, 5.25–5.35, and 5.725–5.825 GHz. These bands are called lower, middle, and upper U-NII bands, respectively. The regulations, which all equipment operating in these bands must satisfy, are in Section 15.401 of the FCC regulations.

Recent regulatory changes

Very recently, important further developments took place. In a Notice of Proposed Rulemaking FCC 03-110 dated June 2003, the FCC amended Section 15.401. To harmonize US regulations with European regulations at the request of the Wi-Fi® Alliance, the FCC allocated additional 255 MHz of spectrum between 5.47 and 5.725 GHz. At the same time, the FCC took steps to protect licensed wireless services operating in this frequency band from possible interference from unlicensed wireless networks. The lower U-NII band is not allowed to be used outdoors, and the maximum EIRP that is allowed for the band of 5.15–5.25 GHz is 200 mW. For the bands of 5.25–5.35 GHz and 5.47–5.725 GHz, the maximum EIRP allowed is 1 W. These bands are shared with radiolocation, Space Research Service, and Earth Exploration Satellite Service. For the 5.725–5.825 GHz band, the maximum EIRP that is allowed is 4 W. These power levels are allowed with antennas that have up to 6 dBi gains. Devices may use antennas with higher gains, but the output power at the transmitter-to-antenna connector must be reduced by the same amount in decibels that the antenna gain exceeds 6 dBi. For fixed point-to-point systems operating in the upper U-NII band, antenna gains up to 23 dBi are allowed with no reduction of output power.

Chapter 1: Introduction

> 💡 FCC regulations are constantly changing. The process for a FCC rule change is based on a petition from a private company, individual, or government organization. Depending on the scope of the intended change, the FCC will issue either a Notice of Inquiry (NOI), or a Notice of Proposed Rulemaking (NPRM), soliciting comments from all interested parties. Any individual, private business, or organization can provide comments and opinions. Once the FCC completes its decision making, the new rule is issued in a Report and Order document. This document not only lists the new rule, but discusses its rationale for adopting it.

Before vendors are allowed to sell products, they must demonstrate that their products comply with the relevant regulations. The FCC plays an important role in certifying all equipment. In the past, to help vendors obtain certification, the FCC issued documents to interpret its own rules. Applicants were required to test their equipment and submit a detailed report to the FCC. A subset of the report is kept in a database open to the public. The FCC interpretation database and grantee database provide useful information to equipment manufacturers. Tradition and common practice are also good guidelines.

Recently there have been notable changes to the role of the FCC in equipment certification. The reason is that the whole certification process used to take a couple of months, which is a significant length in a short product lifetime. Some companies have spent significant amounts of money to ensure that their products comply with the latest FCC regulations, only to find competing noncompliant products on the marketplace. The goals of the changes have been to make the process faster and allow innovative products to reach the market sooner, while still providing adequate protection to existing products and services. To achieve this goal, the FCC has been allowing private industry to develop the technical criteria necessary to protect existing wireless networks and services. The current FCC leadership generally feels that having private industry develop the technical criteria better serves the public interest, rather than having the FCC continue to do so. Ultimately, however, the responsibility rests with the FCC.

> The government regulations in CFR 47 do not answer every question. Furthermore, the rules are not always clear. Some rules even may appear to contradict each other.

The Code of Federal Regulations 47 (CFR 47) [B39] is constantly changing in stride with the development of wireless markets and technologies. On one hand, advances in technology lead to changes in government regulations, and on the other hand, in the process of revising the government regulations the goal is to allow the introduction of new devices and services, without causing an unacceptably high level of interference to existing systems.

Among the recent FCC decisions, in February 2002 the Commission adopted landmark regulations that authorized the limited operation of ultrawide band (UWB) devices in the spectrum above 3.1 GHz. Based on these regulations, UWB has the potential to become a commercially successful technology and therefore there is a significant industry interest.[2]

Future FCC directions

Recently the FCC issued a Notice of Proposed Rulemaking to investigate the feasibility of expanding access for wireless broadband services in the frequency band of 2500–2690 MHz, currently occupied by instructional television. In connection with this initiative, the FCC is soliciting comments on the regulations necessary to promote competition and innovation. This development is considered to be significant because the frequency band is twice as wide as the 2.4 GHz ISM band, which is used by the very successful WLAN and WPAN technologies.

Another policy change contemplated by the FCC is to incorporate performance specifications on the receiver interference immunity. These performance guarantees state that interference is below a certain threshold. While the FCC does not yet plan on issuing regulations subjecting all receivers to mandatory standards, it plans to investigate whether market incentives and voluntary industry programs could be more effective for receiver immunity.

---

[2] In a Memorandum opinion and order issued in 2003, the Commission did not make any changes. In fact, it denied the requests for further restrictions on UWB devices and granted those requests that would not increase the interference of UWB devices, in the FCC's opinion.

Chapter 1: Introduction

In a Notice of Inquiry released in December 2002, the FCC invited comments on the feasibility of expanding the operation of unlicensed devices in the television broadcast spectrum at locations and times when the spectrum is not being used. Unlicensed devices can monitor the frequency spectrum and avoid the use of frequency bands that are used by other devices. The FCC is also exploring whether unlicensed devices should be allowed to operate in other bands such as the 3650–3700 MHz band, provided that they do not cause interference to licensed services. The FCC believes, and rightfully so, that expanding the available spectrum for unlicensed devices will lead to further development of new and innovative types of products for consumers and businesses.

The FCC assumed its present role in 1934, as a result of the 1934 Telecommunications Act of the U.S. Congress. The rationale of having regulations is, arguably, the scarcity of spectrum and potential for interference. However, some modern critics of the FCC argue that scarcity of spectrum is artificial and induced by regulations favoring the broadcast industry [B37]. Currently, there is talk about the FCC withdrawing entirely from the spectrum allocation task. Subsequently, the present spectrum allocation would need to be completely revamped. There are two specific ideas, not entirely in conflict, how can this change be accomplished. The first proposal, based on advances such as dynamic frequency selection (DFS) and transmitter power control (TPC), calls for allowing every intelligent wireless device access to any spectrum [B37]. Devices that cause interference can change their frequency and power of operation. The second proposal calls for all spectrum, like any other resource in the United States, to be available on the market, bought and sold freely by companies, organizations, and individuals, to achieve maximum economic efficiency [B37]. In this case, government organizations, such as defense and law enforcement, could purchase spectrum on the open market for their needs. Clearly, the idea to completely revamp the present spectrum allocations is revolutionary and most likely will not be implemented in the foreseeable future. However, in the long run this idea might become more attractive, especially with further advances in technology.

International Regulations

Canada provides for ISM bands and general operating modes that are almost identical to the United States [B88]. Canadian Radio Standards Specification

RSS-210 provides numerical requirements for narrowband operation in 902–928 MHz and 2400–2483.5 MHz and associated harmonic limits that are identical to FCC 15.249 operation. As of the time of this writing, the Canadian requirements for spread-spectrum operation given in RSS-210 are equivalent to the previous (before May 30, 2002) FCC 15.247 requirements. However, it is likely that Industry Canada [B56] will soon adopt the changes recently made in the U.S. regarding "digital modulation."

The more than 40 member nations of the European Conference of Postal and Telecommunications Administrations (CEPT) have established a fairly high degree of standardization throughout Europe on the operation of low-power radio equipment. Most disagreements concern allowed modes and transmit duty cycles that may be accounted for in software control, allowing the same hardware and technical standards to be used throughout Europe. The European Telecommunications Standard Institute (ETSI) develops technical standards for CEPT countries. Requirements are spread across multiple documents. ETSI EN 300 328-1, 328-2 [B35], [B36] are the European rules for spread spectrum systems, and ETSI EN 300 220-1 provides details on compliance certification [B34]. ETSI 300 328 is in general the formal governing document (though it may be overruled by the regulatory documents of a specific nation). For devices with less than 10 mW EIRP, either the rules of ETSI 300 328 section 5 or the rules of CEPT ERC Recommendation 70-03E Annex 1 may be used. For power levels between 10 mW and 100 mW in the 2400 MHz band, only ETSI 300 328 applies, and direct sequence or frequency-hopping spread spectrum must be used.

 For general radio regulation in Europe, www.ero.dk provides documents and updates, including general description of recommended applications, frequencies, powers, and other specifications.

In Europe, the 433.05–434.79 MHz band segment is the common control and security band. It is primarily limited to these applications because of a general 10% duty cycle limit (see ERC 70-03E, downloadable from www.ero.dk). Europe does not offer an ISM band in the 902–928 MHz range, but it does offer a limited band of 868–870 MHz (see ERC 70.03 [B32] for general rules).

Chapter 1: Introduction

 Important differences in the European rules as compared to the FCC rules are provisions that go beyond preventing interference to other systems—a primary goal of the FCC rules. European rules also attempt to guarantee acceptable system performance.

The Electronic Communications Committee (ECC) within the CEPT is currently studying a number of changes to the short-range device rules. A broad description of some European changes proposed at the time of this writing is captured in [B28]. In Europe at the moment there is an initiative to expand the 868–870 MHz band to cover from 863–870 MHz for nonspecific short-range devices using spread spectrum with power levels up to 25 mW. Some relaxation of the duty cycle limits now applying in the 868–870 MHz range is being contemplated. The changes are generally friendlier to short-range device operation. The use of interference avoidance techniques such as frequency agility, dynamic channel assignment, and listen-before-transmit are encouraged.

The 2400–2483.5 MHz ISM band is also provided for in ERC 70-03 [B32] for short-range devices (any digital modulation form) and also in ETSI 300 328 [B35], [B36] for spread-spectrum devices with data rates equal to or greater than 250 kb/s. Note that European regulations are not unified. There are differences among individual countries.

In Japan, the government has chartered an industry organization called Association of Radio Industries and Businesses (ARIB) [B52] to perform certain quasi-governmental regulatory functions in supporting efficient use of the radio spectrum. The governing document is generally standard ARIB STD-T66 [B7]. This document seems incomplete, especially in the area of how to perform testing to ensure compliance with government regulations. It is also not clear from the standard exactly which specifications carry the force of law and which are recommendations. DSSS operation in the 2.4 GHz band is allowed with maximum output power of 10 mW/MHz, equivalent to 10 dBm/MHz. a minimum spreading factor of 5 is required. FH is allowed with maximum output power 3 mW/MHz or 4.77 dBm/MHz, and non–spread spectrum with 10 mW/MHz. Other requirements given in this standard include a maximum antenna gain of 2.14 dB at full power of 10 mW/MHz, though higher gain antennas can be used with a commensurate decrease in maximum transmit power.

Chapter 1: Introduction

In summary, the 2400–2483.5 MHz band is the only almost worldwide allocation of spectrum for unlicensed usage without any limitations on applications and transmit duty cycle. It provides up to 1 W transmit power in spread spectrum modes in the U.S., up to 100 mW in Europe, and up to 10 mW/MHz in Japan.

Note that spectrum allocation is only one part of the regulations. Spurious suppression requirements vary considerably not only by nation, but also by test methodology and the use of averaging. Because they are lower in frequency and thus accessible in lower capability integrated circuit processes, the European 868–870 MHz and the U.S. 902–928 MHz bands are special cases of regional bands that are still useful even though they are not allocated worldwide. Given that antennas remain approximately omnidirectional, they also provide larger antenna aperture than 2400 MHz. So at a given transmit power, data rate, receiver sensitivity, and reliability level, they will provide greater range.

Unfortunately, government regulations around the world are not identical (see Table 1–1). This creates problems especially in the area of wireless data communications, where devices can easily cross international boundaries. Users should make an effort to configure devices so that appropriate government regulations are always observed. Regulatory bodies around the world generally recognize that it is desirable for regulations to be identical and whenever possible make an effort to achieve this. Unfortunately, it does not seem possible to achieve complete harmonization of the regulations, because this would require the costly relocation of some existing wireless services to different frequency bands.

Table 1–1: Maximum output power in the ISM bands

Frequency band	Region	Maximum output power
2400–2483.5 MHz	USA	1000 mW
	Canada	1000 mW (with some limits on installation location)
	Japan	10 mW/MHz
	Europe	100 mW EIRP or 10 mW/MHz
902–928 MHz	USA and Canada	1000 mW
868–868.6 MHz	Europe	25 mW

STANDARDIZATION BODIES

The development of large markets for wireless communication services and equipment depends not only on the availability of sufficient spectrum in appropriate frequency bands, but also on the availability of standards. The main reason that large markets are not possible without standards is that big telecommunication service providers want to have the choice of buying equipment from multiple suppliers, rather than having to purchase equipment from a single company. In the absence of a standard, big service providers simply refrain from using the technology. The existence of standards also helps small technology companies to enter large markets by reducing risks associated with providing products such as semiconductor chipsets, software, firmware, etc. Standards can be viewed as voluntary agreements among technology and semiconductor companies, equipment manufacturers, and service providers. Until several years ago, there were no standards for wireless data communication; the total market was small, fragmented, and dominated by several proprietary technologies. The dramatic market growth is a direct result of the emergence of several standards, such as IEEE 802.11a, 802.11b, and 802.11g™. Standards allow equipment from different vendors to work together in a network. Standards create mass markets for equipment, which creates economy of scale for manufacturers. The economic effect of standards is even more significant than simply enabling mass markets. The market for standard-compliant products is characterized by significant competition, which results in lower prices. To counter this trend, many vendors are looking for ways to differentiate themselves in the marketplace, while still offering standard-compliant products. Overall, standards significantly stimulate creativity, innovation, and entrepreneurship.

Because standards enable mass markets, intellectual property required for the implementation of a standard is highly valuable. Accordingly, some of the most valuable patent claims are claims that cover a standard or parts of a standard. Therefore, the intellectual property policy of the standards-making body is very important. In industry consortia, the intellectual property behind their specification may or may not be available for licensing by companies that are not a member of the relevant consortium. This means that non-member companies that want to manufacture and sell products complying with a particular consortium's specification may have to pay license fees, which can create a barrier to market entry. In some cases, the intellectual property may not be available for licensing at all. When the intellectual property is available, the conditions under which it can be licensed may create other barriers to market entry. For example, intellectual property can be made available on a reciprocal basis. The specification reflects the intellectual property position and the business models of the members of the consortia. Specifications produced by industry consortia are intended to maximize the revenue of the members of the consortium. The Bluetooth® SIG and HomeRF are industry consortiums that have recently produced specifications for wireless data communication.

With the standards revolution of the last five years came a plethora of standards-making bodies. In general, standards-making bodies are of two types: industry consortium and open organizations. Industry consortium type bodies are inherently closed. The companies that establish the consortium dictate the rules of the standards-making process. Membership in the consortium may or may not be open.

Open organizations, such as standards-development organizations, are designed to develop standards that do not favor any one company, but rather aim to benefit the public good. The IEEE is an open standardization body. Before an IEEE standard is adopted, companies that are known to have essential patents are asked to provide a patent letter of assurance stating that they will either not enforce any of their present or future patents that are required (essential) to implement either mandatory or optional portions of the standard against anyone wishing to comply with the standard, or that they will provide licenses, either without compensation or at reasonable rates, on a nondiscriminatory basis, with reasonable terms and conditions to those seeking to implement the standard. The meaning of "reasonable" is somewhat loosely defined. There is no guarantee from the IEEE that a standard does not violate the intellectual property of other parties, including parties unwilling to license their intellectual property on reasonable terms. Furthermore, only a court of law can accurately determine the meaning of "reasonable" and "nondiscriminatory."

Chapter 1: Introduction

Within the IEEE, standards for data communications are developed by IEEE 802 community, called the Local and Metropolitan Area Networks Standards Committee (LMSC) and sponsored by the IEEE Computer Society [B57]. IEEE 802 has the charter to develop and maintain global standards and recommended practices for computer communication. Some successful IEEE 802 standards are the IEEE 802.3™ or Ethernet standards [B91], IEEE 802.5™ or Token Ring standards and the IEEE 802.11™ or Wi-Fi standards. They all have been adopted by the ISO/IEC Joint Technical Committee 1 (JTC1) as International standards.

IEEE 802 was formed in February 1980 and has held three plenary meetings a year since then. It is legend that the 802 number was chosen to reflect the year and month it was created, however, this is untrue. The number 802 was simply the next number in the queue.

Although IEEE is rooted in the United States, it enjoys strong international participation at the IEEE meetings, and many standards produced by the IEEE are international standards. IEEE 802 consists of several working groups, which are organized around significant applications. The IEEE 802 standards deal with the physical and data link layers in the ISO open systems interconnection (OSI) reference model. IEEE 802 standards specify the data link layer in two sublayers, logical link control (LLC) and medium access control (MAC). The LLC is standardized in IEEE Std 802.2™ and is common to all IEEE 802 MACs.

Initially, work on wireless data communication took place in Working Group IEEE 802.4™, which worked on token-passing multiple-accessing method. In 1990 the Working Group reached the conclusion that token passing is not a suitable protocol for wireless local area networks (WLAN), and the IEEE 802.11 Working Group was established [B58].

Over the years other MAC and PHY working groups have been established. Some of these working groups did not publish a standard, and other groups became inactive after producing a standard. While not all standards have been successful, it must be noted that at present IEEE Std 802.11™ is the most commercially successful wireless standard in the world.

In 1999 two other wireless working groups were established—IEEE 802.15 [B59] devoted to WPAN and IEEE 802.16 [B60] for wireless broadband access. Membership in these working groups is by individual only and is open to anyone.

Company membership is not allowed. Only individuals vote. Membership includes representatives mainly from the United States, Canada, Europe, Japan, and Australia. Standards established by the working group first become IEEE standards, and then are submitted to the ISO/IEC for adoption as international standards. The commercial success of IEEE 802.11 and IEEE 802.15 and the work in IEEE 802.16 has made IEEE 802 recognized internationally as the main wireless standards-making body.

Just because an international standard exists does not mean that it will succeed in the marketplace. The success of a standard depends not only on the quality of the technology, but also on business and political reasons. Therefore, there has to a be certain alignment among interested companies on these business and political issues on which the success of the standard depends. In an industry consortium, this alignment is easier to achieve. To achieve this alignment for a standard produced by an open standardization body, companies often form other industrial organizations. Such organizations relevant to IEEE 802.11 and IEEE 802.16 are Wi-Fi [B61] and WiMAX™, respectively. The tasks of these organizations include testing to certify interoperability among products from different vendors and promoting the standard-compliant products in the marketplace.

IEEE 802.11, the Working Group for Wireless Local Area Networks (WLANs), is responsible for developing Carrier Sense Multiple Access/Collision Avoidance (CSMA/CA)–based WLAN standards within IEEE 802. Since IEEE 802.11 was formed in July 1990, it has produced the ISO/IEC 8802-11:1999 (IEEE Std 802.11-1999) standard [B68] with several amendments (once referred to as "supplements"). Amendment IEEE 802.11b defines a physical layer achieving data rates of 11 megabits per second (11 Mb/s) in the 2400–2483.5 MHz ISM band [B70]. The amendment IEEE 802.11a is another physical layer for operation between 6 Mb/s and 54 Mb/s in the unlicensed bands above 5 GHz [B69]. IEEE 802.11b and IEEE 802.11a were adopted in 1999. Recently, however, the IEEE 802.11 Working Group has been working on several other amendments.

IEEE 802.11F™[B72], approved in 2003, is an amendment to the medium-access control layer that enables complex wireless networks to be built. IEEE 802.11e™ is another MAC amendment, which would allow higher Quality of Service [B71]. IEEE P802.11i™ is an amendment to ensure security [B75]. These amendments faced several technical challenges and will likely not be completed in 2003.

IEEE 802.11h™ is an amendment to IEEE 802.11a, which was completed in 2003 [B74]. IEEE 802.11g™ is yet another physical layer for data rates over 20 Mb/s in the 2400–2483.5 MHz band [B73]. It was also completed in 2003. Table 1–2 provides a summary of the completed and active projects within 802.11.

Table 1–2: 802.11 amendments

Number	Description
802.11a	Physical layer for the 5 GHz UNII bands, 6–54 Mb/s
802.11b	Physical layer for the 2.4 GHz ISM band, 5.5 and 11 Mb/s
802.11c	Supplement to support MAC bridge operation
802.11d	Specification for operation in different regulatory domains
802.11e	Enhancements for QoS (active)
802.11F	Interaccess point protocol
802.11g	Physical layer for operation in the 2.4 GHz ISM band
802.11h	Spectrum and power management enhancements to 802.11a
802.11i	Security enhancements (active)
802.11j	Enhancement to 802.11a for operation in 4.9–5.0 GHz in Japan (active)
802.11k	Radio resource management (active)
802.11m	Technical corrections and clarifications (active)
802.11n	High-throughput enhancements (active)

IEEE 802.15, the standards Working Group for Wireless Personal Area Networks, is responsible for developing standards for short distance wireless networks. IEEE 802.15 was formed in July 1999. The group has four projects and correspondingly four task groups. the first task group, IEEE 802.15.1™, produced a WPAN standard in the 2.4 GHz ISM band with PHY and MAC layers equivalent to the PHY and the MAC layers of Bluetooth [B78]. The success of both Bluetooth and

IEEE 802.11, which operate in the 2.4 ISM GHz, created interference problems between the two standards. Within IEEE 802.15, Task Group 2 was established to develop guidelines on how these two standards can coexist [B79]. The desire to achieve higher data rates led to the formation of Task Group 3 [B80]. The desire to develop WPAN devices with extended functionality and low data rates led to the formation of Task Group 4 [B81]. Table 1–3 summarizes the 802.15 projects.

Table 1–3: 802.15 projects

Number	Description
802.15.1	WPAN based on portions of Bluetooth v.1.1
802.15.1a	WPAN based on portions of Bluetooth v 1.2
802.15.2	Coexistence of WPAN with other systems in the 2.4 GHz band
802.15.3	High-rate WPAN
802.15.3a	Additional UWB physical layer for 802.15.3
802.15.4	Low-rate WPAN
802.15.4a	Additional physical layer for 802.15.4

IEEE 802.16, the Standards Working Group, is responsible for developing standards and recommended practices for Broadband Wireless Access Networks. IEEE 802.16 was formed in March 1999 [B60]. The group has several projects:

a) Air Interface (including a MAC and a PHY) for operation between 10 and 66 GHz

b) IEEE 802.16a™, an amendment specifying additional physical layers and appropriate MAC modifications for operation between 2 and 11 GHz, including licensed and unlicensed bands

c) IEEE Std 802.16.2™ and 802.16.2a™, which provide a recommended practice for coexistence

d) IEEE 802.16e, being developed at present to support mobile subscribers

In addition IEEE 802.16 has developed interoperability documentation, which is normally done outside of IEEE 802 standards (see Table 1–4).

Table 1–4: 802.16 projects

Number	Description
802.16	Fixed broadband wireless systems between 10 and 66 GHz
802.16a	Amendment for operation between 2 and 11 GHz
802.16c	Enhancement including system profiles between 10 and 66 GHz
802.16.2	Coexistence between 10 and 66 GHz
802.16/Conf01 802.16/Conf02 802.16/Conf03	Test and conformance specifications (active)
802.16d	System profiles (active)
802.16e	Enhancement to support mobility (active)

Other related working groups have been established recently. The IEEE 802 standards committee decided that the monitoring and active participation in radio regulatory activities around the world are important and established a Radio Regulatory Technical Advisory Group—IEEE 802.18. One of the first priorities of IEEE 802.18 is to monitor the global allocation of a significant part of the 5 GHz band

As the number of wireless standards increased, the issues of coexistence among the different wireless standards became prominent. IEEE 802.19 is a Coexistence Technical Advisory Group (TAG), established in 2002. The responsibilities of IEEE 802.19 include helping other working groups such as IEEE 802.11, IEEE 802.15, and IEEE 802.16 address coexistence issues. IEEE 802.19 may develop coexistence documentation of its own.

Chapter 1: Introduction

> 💡 Not all standards are commercially successful. Usually the lack of success is due to business reasons, rather than technology. At present, IEEE 802.11 and Bluetooth (IEEE 802.15.1) are the most successful wireless standards.

IEEE 802.20 is yet another recent working group. Its mission is to develop the MAC and PHY layers for mobile broadband wireless access systems. IEEE 802.20 was established in December 2002 and at present is in the process of accepting proposals.

The European Telecommunications Standardization Institute (ETSI) is another open standardization organization [B54]. Membership in ETSI is open only to companies and organizations, and not to individuals. Membership is open to companies from the entire world, and many American and Japanese companies participate. ETSI has produced two families of wireless standards called HiperLAN and HiperAccess, which are technologies for wireless LAN and wireless broadband access, much like IEEE 802.11 and IEEE 802.16.

On an international scale, a major international standardization body is the International Telecommunications Union (ITU), an agency of the United Nations. The ITU is responsible for communications standards and for treaty-based agreements on spectrum management. Its specific role in spectrum management is to minimize radio interference by establishing international rules standardizing the use of radio-frequency bands. It organizes and administers World Radio Congresses, which are held regularly to update the radio regulations.

WIRELESS CHANNELS

The first step in the design of a radio network is the characterization of the radio channel, because the performance of a wireless network—signal coverage and achievable data rate—heavily depends on the wireless communication channel. To help determine signal coverage and achievable data rate, a link budget calculation is performed. A block diagram of a radio system is shown in Figure 1–1.

Chapter 1: Introduction

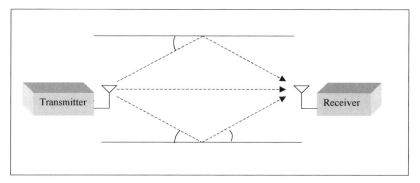

Figure 1–1: A general block diagram of a wireless transmitter and receiver

The transmitting antenna converts the radio-frequency (RF) signal into an electromagnetic wave. The effective isotropic radiated power (EIRP) is equal to the transmitted output power minus cable loss plus the gain the transmitting antenna:

$$\text{EIRP} = P_{out} + C_t + G_t \qquad \text{Eq. 1–2}$$

where P_{out} is transmitted output power in dBm, C_t is the cable loss in dB, and G_t is the gain of the transmitting antenna in dBi. The gain of an antenna is a measure of its directivity. It is the ratio of the radiation intensity in a given direction over the radiation intensity of an isotropic antenna. Isotropic antennas have gain of 0 dBi and radiate equally in all directions. The transmitted electromagnetic wave propagates in the wireless medium. It is intuitively clear that only a portion of the transmitted electromagnetic wave's energy will reach the receiver antenna. The most basic model of RF propagation is known as the "free space model." The free space propagation model is used to predict received signal strength when the transmitter and receiver have a line of sight path between them. The power of such a transmitted signal would degrade with an inverse square rule, or 20 decibels (dB) per decade. More specifically the received power will be given by the Friis free space equation:

$$P_r = \frac{P_t G_t G_r \lambda^2}{C_t C_r (4\pi)^2 d^n L} \qquad \text{Eq. 1–3}$$

where P_r and P_t are the received and transmitted power, respectively; G_t, G_r are the transmitter and receiver antenna gains; C_t and C_r are the cable losses at the transmitter and receiver; d is the distance between transmitter and receive;, and n is a number that depends on the wireless environment. In indoor environments, n can vary from 1.6 to 3.3, where values smaller than 2 can be achieved in corridors. The wavelength is $\lambda = c/f$, where c is the speed of light and f is the frequency of operation. The wavelength is 12.5 cm and 5.2 cm, correspondingly for the 2.4 GHz and 5.8 GHz bands. L represents a signal loss factor, including transmission losses, filter losses, and antenna losses. $L = 1$ signifies no losses. Assuming no cable losses, the free space path loss equation can be written as

$$P_r = \frac{P_0}{d^n} \qquad \text{Eq. 1-4}$$

where P_0 is the received power at a distance of 1 m and is equal to

$$P_0 = \frac{P_t G_t G_r \lambda^2}{(4\pi)^2 L} \qquad \text{Eq. 1-5}$$

Measured in decibels, we have

$$P_r = P_0 - 10n\log d \qquad \text{Eq. 1-6}$$

Assuming that $L = 1$, the power loss at 1 m is approximately 40 dB for 2.4 GHz and 47.5 dB for 5.7 GHz. In free space, n is equal to 2, producing $P_r = P_0 - 20\log d$, which explains where the 20 dB-per-decade rule comes from.

The free-space path loss model is valid for distances that are in the far field of the transmitting antenna, i.e., distances that exceed $2D^2/\lambda$, where D is the largest physical linear dimension of the antenna. Note also that the path loss is random according to a Gaussian (or normal) distribution, and the model provides only the average value. This effect is called log-normal shadowing. The standard deviation of this Gaussian distribution in the frequency bands of interest here is 8 dB for residential indoor applications and between 10 and 12 dB for indoor office applications [B122].

As an example, if the output power P_t is 30 dBm (or equivalently, 1000 mW) and the transmit and receive antennas have gains of –5 dBi each, and if the cable losses in the cables to the antennas are 1 dB, the EIRP is 30 – 5 – 1 = 24 dBm. For a transmission at 2.4 GHz, the loss at a distance of 3 km is

$$10\log\frac{0.125^2}{(4\pi)^2 3000^2} = 110 \text{dB}$$

(assuming $L = 1$), and the received power would be $P_r = 24 - 110 - 5 - 1 = -92$ dBm. In this example, a receiver with sensitivity of at least –92 dBm is required.

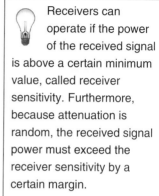

Receivers can operate if the power of the received signal is above a certain minimum value, called receiver sensitivity. Furthermore, because attenuation is random, the received signal power must exceed the receiver sensitivity by a certain margin.

However, the performance of a wireless communication system is governed by two groups of factors. The first factor is the received power, determined by the transmitted power and the gains and losses on the link. The second factor includes the signal processing used to encode and modulate the signal. The power of the received signal cannot characterize the quality of the received signal. A measure of the quality of the received signal is the signal-to-noise ratio (SNR):

$$\text{SNR} = \frac{P_r}{P_n} \qquad \text{Eq. 1–7}$$

where the noise power is $P_n = KT_{sys}B$, $K = 1.38 \times 10^{-23}$. J/K is Boltzmann's constant and B is the signal bandwidth in Hertz. T_{sys} measured in Kelvin (K) is the equivalent system noise temperature which represents all noise sources: external electromagnetic noise picked up by the antenna, noise generated by the antenna due to thermal emission by the antenna material, and noise generated internally by the receiver.[3] The external noise can be naturally occurring and unavoidable like atmospheric noise or man-made, like the interference from other wireless systems operating in the same frequency band. The biggest source of unavoidable noise is the thermal noise generated in the receiver electronics. The finite precision of the arithmetic operations in the receiver is another noise source. Because at the

---

[3] If T_{sys} is assumed to be 289 K, the noise power spectral density will be –204 dB/Hz. In addition, practical receivers will generate noise between 5 and 12 dB, and the finite-precision arithmetic operations will lead to an additional implementation loss of typically about 5 dB.

receiver the SNR must be above a certain threshold SNR_{th}, the path loss will limit the signal bandwidth (and therefore the data rate) according to

$$B \leq \frac{P_t G_t G_r \lambda^2}{SNR_{th}(4\pi)^2 L N_0 d^n}$$
Eq. 1–8

or the signal coverage,

$$d \leq \left[\frac{P_t G_t G_r \lambda^2}{SNR_{th}(4\pi)^2 L N_0 B}\right]^{1/n}$$
Eq. 1–9

where N_0 is the one-sided noise power spectral density.

This free-space model is accurate when there is a line of sight (LOS) between the transmitter and receiver. The line-of-sight condition can be more precisely defined in the following way. The optical LOS is the imaginary straight line between the two antennas. In wireless communications, it can be assumed that a clear LOS exists if a certain area around the optical LOS is free of obstacles. This area is called the Fresnel zone, and its radius is $R = \frac{1}{2}\sqrt{\lambda d}$. When the Fresnel zone is not clear of obstacles, there is not a line of sight, and the additional obstructions in the signal path will cause additional signal loss. The amount of additional signal loss will be random.

Table 1–5 shows the attenuation that some building materials can cause. When the signal path is known, all obstacles with their attenuation contributions can be added to the equation for a total path loss figure. It is this equation that is used when determining the signal strength at a particular point in many of the propagation models.

In addition to attenuation, the transmitted electromagnetic wave in Figure 1–1 will be reflected by walls, ceilings, and other objects if indoors, or buildings and terrain features if outdoors. Therefore, the signal will reach the receiver via multiple paths. This phenomenon is called multipath. If LOS is present, the LOS component arrives first at the receiver and is the strongest. The transmission delay of the LOS component is $\tau = d/c$, which is approximately $3d$ ns, or 3 ns delay per meter. In the absence of LOS, the signal on the first arriving path may not be the strongest. When the signal is reflected from ceilings, walls, etc., part of the signal energy is reflected and part of the signal energy will start propagating through the ceilings, walls, etc. Therefore, the signals on the different paths will have

Table 1–5: The average additional attenuation of some building materials at 2.4 GHz

Material	Additional Loss(dB)
Open space	0
Window	3
Floor/ceiling	12–15
Wall (wood)	10
Wall (metal and concrete)	15–20

different strengths. Because these paths are of different lengths, the arrival times of the signals on the different paths will be different, as shown in Figure 1–2.

The difference between the arrival times of the first and last paths is called delay spread. Each multipath component is characterized by amplitude and phase. The phases can be accurately modeled with uniform distribution between 0 and 360 degrees. The amplitudes have exponentially decaying Rayleigh distribution. The multipath phenomenon leads to fading—fluctuation in the received signal.

Fading is the result of the randomness with which the multipath components add at the receiver. Fading depends on both the spatial position of the receiver and on the frequency of transmission.

Clearly, for some positions of the receiver, the multipath components will arrive with opposite phases and will cancel each other. At these positions, the receiver is in a null, and communication is not possible. Fading also depends on the frequency of transmission. For some frequencies, the multipath components will add destructively, resulting in deep fades or spectral nulls. The complex impulse response of the multipath channel can be represented as in:

$$\sigma_k^2 = \sigma_0^2 e^{-kT_s/T_{RMS}} \qquad \text{Eq. 1–10}$$

$$\sigma_0^2 = 1 - e^{-kT_s/T_{RMS}} \qquad \text{Eq. 1–11}$$

where $N\left(0, \frac{1}{2}\sigma_k^2\right)$ is a zero mean Gaussian random variable with variance $\frac{1}{2}\sigma_k^2$, and $\sigma_0^2 = 1 - e^{-kT_s/T_{RMS}}$ is chosen in such a way so that the condition $\Sigma\sigma_k^2 = 1$ is satisfied. The number of samples to be taken in the impulse response should ensure sufficient decay of the impulse response tail—for example, $k_{max} = 10 T_{RMS}/T_s$. The radio channel is linear, and therefore is completely described by its impulse response. The goal of channel modeling is to estimate the impulse response. Note that the impulse response varies with the relative position of the receiver with respect to the transmitter and also varies with time.

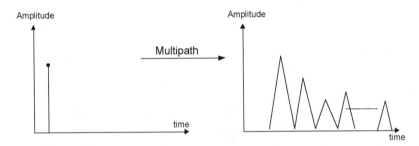

Figure 1–2: A transmitted pulse as a result of the multipath effect will appear to the receiver as multiple smeared impulses

When multiple pulses (or symbols) are transmitted they will start overlapping in time when they reach the receiver (Figure 1–3). In other words, the symbols will interfere with each other, or equivalently there will be intersymbol interference (ISI), as shown in Figure 1–3.

The root mean square (rms) delay spread is often used to characterize wireless channels. The rms delay spread and the symbol period are used to estimate the amount of ISI caused by multipath channel. Because the distances that the individual paths travel are smaller indoors, the rms delay spread for indoor communications is smaller than outdoors. The delay spread tends to increase as the distance between the antennas increases, in which case the path loss increases as well.

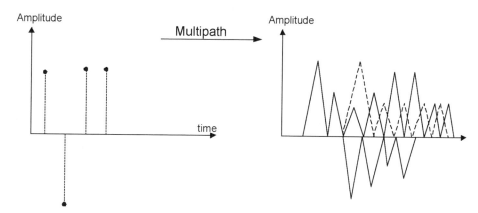

Figure 1–3: Multiple transmitted impulses as a result of the multipath will be received as shown on the right

The characteristics of reflected paths between the transmitter and receiver depends also on polarization, incidence angle, and the material's complex permittivity. Therefore, polarization and antenna radiation pattern can significantly affect indoor propagation characteristics. Directional antennas can reduce the rms delay spread compared with omnidirectional antennas. Circular

 Typical values for the rms delay spread are 50 ns in a home environment, between 100 ns and 150 ns for offices, and about 250 ns for large indoor spaces such as warehouses, airports, shopping malls, and convention centers.

polarization also can reduce the rms delay spread compared with linear polarization, independent of the frequency of operation. It is finally noted that reducing the rms delay spread is not always desirable, because some communication systems specifically take advantage of multipath effects.

Wireless communication channels clearly change with time as well. This change in general is a result of moving vehicles, objects, people, etc., as well as transmitter and/or receiver mobility. Mobility will lead to an effect called Doppler spread. Doppler spread is a measure of the spectral broadening caused by the channel time variation. The Doppler frequency is $f_D \leq \frac{v}{\lambda}$. For example, for operation at 5 GHz, mobility of 5 mph will lead to a Doppler frequency of 37 Hz, meaning that the channel will change at a rate of 37 Hz. Mobility of 60 mph will result in a Doppler frequency of 444 Hz. However, when using high-speed wireless data communication products, users require limited mobility. As a result, it is both convenient and realistic to assume that the channel characteristics do not change within a packet, only from packet to packet. In the case of a laptop computer with a wireless network interface card, the LOS path can be shadowed by people moving in the vicinity of the laptop. Some measurements have observed fades of 10–15 dB in these cases [B122].

Multipath fading together with attenuation and noise limits the achievable signal coverage and data rate in high-speed wireless data communication. To mitigate the channel conditions, wireless standards use the following three methods:

a) Power management

b) Signal processing

c) Antenna diversity

Mobile and portable devices must be battery-powered. Therefore, low power consumption is crucial. The wireless data communication standards use sophisticated power management to reduce power consumption as much as possible. Receivers are generally implemented with equalizers to compensate for ISI caused by the multipath channel. Multicarrier modulation is another technology, very suitable for multipath channels. Advanced antenna systems are used in particular by broadband wireless access systems.

Wireless channels are described in considerably greater detail in [B116] and [B121].

INTRODUCTION TO CRYPTOGRAPHIC ALGORITHMS

The goal in this section is to introduce the specific cryptographic algorithms that are used in the IEEE 802 standards. Cryptographic algorithms are discussed in greater detail in [B132].

In secret-key cryptographic system, the sender uses a secret key for the encryption. The intended receiver recovers the plaintext using the same key. It is the sharing of a secret that makes the communication secure. This method can be used in authentication. Suppose that two devices want to authenticate each other using this approach. Then one device can send a random number (the challenge) to the other device. The other device will respond with an encrypted version of the random number. The receiver will decrypt the number, and if the result is equal to the original random number, the authentication procedure is considered successful. If mutual authentication is desired, then the second device can issue its own challenge by sending its own random number. It is extremely important that the random numbers used in each challenge be different. Otherwise an eavesdropper can determine the key that is shared by observing the challenges and responses. The term nonce (number once) is used to describe the random numbers.

 A message to be encrypted is called plaintext and the resulting encrypted message is called ciphertext. The objective of encryption is to be practically impossible for an unauthorized party to understand the contents of the ciphertext. The main types of cryptographic systems are secret-key and public-key systems.

In public-key cryptography keys are not being shared. It relies on two different keys, a public key and a private key. Encryption is performed using a public key, and decryption is performed by using a private key. Privacy is provided, because only the holder of the specific private key can decrypt the message. Public-key systems can also be used for authentication. In this case, one device selects a nonce, encrypts it using the other device's public key, and transmits it to the other device. The other device decrypts the received message with its private key and responds with the originally selected nonce.

The Data Encryption Standard (DES) was adopted in 1977 by the National Bureau of Standards, now National Institute of Standards and Technology (NIST). DES is a shared-key cryptographic algorithm. In the encryption process, DES first divides the data into blocks of 64 bits. Each block is separately encrypted into a block of 64-bit ciphertext. The key is 56 bits. The encryption algorithm has 19 steps. Each step has a 64-bit input and produces a 64-bit output. The first step is a permutation independent of the key. The last step is a permutation that is the inverse of the initial permutation. The stage immediately before the last simply swaps the 32 bits on the left with the 32 bits on the right. Each of the remaining 16 iterations performs the same processing, but uses a different key. The key at every step is generated from the key at the previous step by applying a permutation, circular rotation, and another permutation. Out of the 56 bits, only 48 bits are used for the key at every iteration.

At each iteration the following operation is performed: First, the 64-bit input is divided into two equal portions, L_{i-1} and R_{i-1}. The output L_i is equal to R_{i-1}. The right part of the output R_i is the result of a bitwise XOR of the left part of the input and a function of the key at the given iteration K_i, and the right part of the input, $R_i = L_{i-1} \oplus f(R_{i-1}, Ki)$. This is the electronic codebook (ECB) mode of DES.

ECB mode may not be secure when the structure of a message is known to an attacker, as is typical in communication standards. Improvements to ECB are possible. One way to improve the algorithm is at every step to XOR the current plaintext block with the preceding ciphertext block. The first plaintext block is XORed with an initialization vector (IV). This is the cipher block-chaining (CBC) mode. Another problem with DES is that the 56-bit key is not sufficiently long for protection against brute-force attacks. An improvement called triple DES mitigates this problem. Triple DES uses two keys and therefore the total key length becomes 112 bits. According to triple DES, ciphertext C is produced from plaintext P by

$$C = E_{K1}(D_{K2}(E_{K1}(P)))$$ Eq. 1–12

The corresponding decryption algorithm is

$$P = D_{K1}(E_{K2}(D_{K1}(C)))$$ Eq. 1–13

If the two keys are identical, triple DES reduces to single DES.

In 1997 NIST announced a public contest to select the successor of DES. In 2001 the Rijndael proposal by Belgian scientists Rijmen and Daemen was selected as the Advanced Encryption Standard (AES). The Rijndael algorithm encrypts 128-bit blocks of data with keys of 128, 192, and 256 bits long. It can be implemented in a very efficient manner.

While DES and Rijndael are shared-key algorithms, the RSA algorithm is the most popular public-key algorithm. It is named after its inventors Rivest, Shamir, and Adleman. It is based on the fact that while it is simple to find the product of two numbers, factorization is more complex. The public and private keys are generated first by choosing two large prime numbers p and q and multiplying them ($pq = n$). Then, a number e is found, which is relatively prime to $(p-1)(q-1)$, and a number d is found so that $de = 1 \bmod (p-1)(q-1)$. The public key is $\{e,n\}$, and the private key is $\{d,n\}$. This algorithm guarantees that for every number P smaller than n, $P^{de} = P \bmod n$. The encryption operation is $C = P^e \bmod n$. The decryption operation is $C^d = P \bmod n$.

Secret-key algorithms are computationally simpler. However, key management is complex. Public key algorithms are computationally intensive. Public-key algorithms simplify key management by requiring each user to have a secret private key and a public key that can be freely distributed. In practice, public-key algorithms require certificates to verify that a given public key corresponds to a certain user.

DESIGN OF A WIRELESS COMMUNICATION STANDARD

In the early days of computer networking, different vendors used systems with incompatible architectures. This required customers to stay with a single vendor. To remedy this situation, the International Standardization Organization (ISO) developed a seven-layer model for communication systems. This model is shown in Figure 1–4. The physical layer deals with the transfer of bits over an actual communication channel. The data link layer deals with the transfer of frames. The data link layer includes framing and address information, as well as flow control. The network layer provides for the transfer of packets across a communication network. It deals with finding a path in the network (routing) and handles congestion from temporary increases in traffic. The transport layer deals with the transfer of complete messages. The session layer controls the data exchange. The presentation layer handles the conversion of data from a machine-dependent form to a machine-independent form, and back. Finally, the purpose of the application layer is to provide services to the various applications.

The IEEE 802 standards deal with the lowest two layers in the ISO open systems interconnection (OSI) reference model: the physical and the data link layer

Chapter 1: Introduction

(Figure 1–4). Furthermore, IEEE 802 specifies the data link layer in two sublayers, logical link control (LLC) and medium access control (MAC). The LLC is situated above the MAC layer. It is standardized in IEEE 802.2 and is common to all IEEE 802 standards. The LLC provides two services to the next higher layer. These are the LLC data service and the LLC management service. The features of the LLC layer are packet segmentation and handshake. The network layer provides functionality to configure and maintain the topology of the network and to interface with the application through the API.

The different wireless standards developed by IEEE 802, like the other IEEE 802 standards, specify the MAC layer and the physical layer. In this section, the general requirements of these two layers are analyzed.

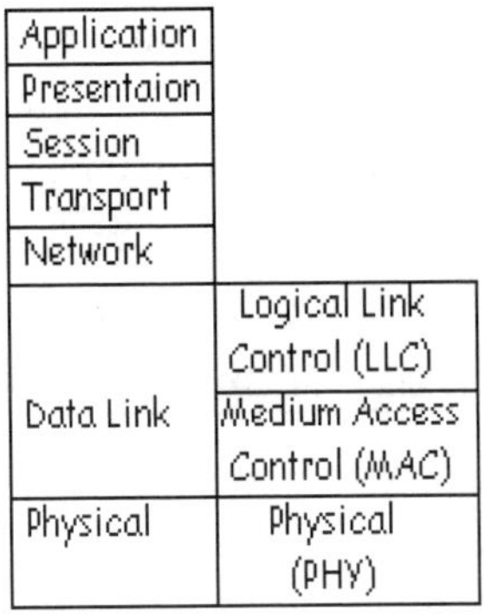

Application	
Presentaion	
Session	
Transport	
Network	
Data Link	Logical Link Control (LLC)
	Medium Access Control (MAC)
Physical	Physical (PHY)

Figure 1–4: Mapping of ISO OSI to IEEE 802

MAC requirements

The access protocol defines the way the medium (the communication channel) is shared among multiple users. The medium in which wireless devices operate is a good example of a shared medium; here the medium is the space through which the radio waves propagate. The ultimate goal of a MAC for wireless communication is to allow a large group of otherwise uncoordinated users to efficiently use this shared medium. Therefore, the choice of a MAC protocol depends on the nature of the traffic and the performance demands of the users.

Traffic can be classified in two groups: periodic and bursty. When the interarrival variance between messages is very small, traffic is said to be periodic.[4] Signals such as voice and video generate periodic traffic. Periodic traffic requires a limit on the maximum end-to-end delay and the delay variation (jitter). Because the delay depends mainly on the time to assign channel access, the MAC design has a big influence on the signal delay. The data rate of a periodic traffic source is approximately constant. Therefore, the use of a dedicated, circuit-switched connection for periodic traffic is justified. Bursty traffic is characterized by messages of arbitrary length separated by intervals of random duration. Data communication (e.g., e-mail, Internet access, file sharing) in an office environment is an example of bursty traffic. The delay and jitter are not important for bursty traffic. A bursty traffic source leads to a data rate that varies considerably. The peak data rate is much higher than the average data rate. If enough capacity over a dedicated connection is provided to satisfy the peak demand channel, utilization will be poor. Packet-switched connections perform much better for bursty traffic. Because communication systems must support not only Internet access, but also voice and video, both traffic types must be handled. This requirement makes the design of a MAC a difficult task. In addition, in wireless networks, the wireless channel is also the only means to coordinate the stations in the network.

The law of large numbers applied to networking means that the combined requirements of a large number of users is equal to the sum of the average requirements of each user. In other words, although the peak demand for each user can be high, when there are many users, it is the average demand that matters because only a fraction of all users have data to transmit at any given time. Still, if several users attempt transmission simultaneously, the result will be a collision. The MAC must resolve these access contentions.

In general, access methods fall into three general categories: contention methods, polling methods, and time-division multiple-access (TDMA) methods. Contention protocols are also called carrier-sense multiple access (CSMA) or listen-before-talk. If a station has data to transmit, it senses the channel for a certain time period before transmitting. For each station, the length of this time period is random within a predefined interval. If the channel has been free during this entire time

---

[4] Note that this does not mean strictly periodic in the mathematical sense.

period, the station transmits. Otherwise, the station defers transmission and waits before again sensing the channel. In this way, the probability of collisions on the wireless medium is minimized, but is not zero. If two stations that have data to transmit choose time periods of equal lengths, collision will occur. In a CSMA with collision detection (CD), which is used in wired Ethernet (IEEE 802.3), the transmitting stations stop the transmission when a collision is detected. An acknowledgment is required for every transmission. Packets that are not acknowledged, due to collision or unsuccessful reception, are retransmitted. Retransmissions reduce the throughput. Under heavy traffic, the end-to-end delay performance of the network becomes poor because there are many collisions and retransmissions. The success of Ethernet demonstrated that contention systems are well suited to bursty traffic. However, contention systems have one fundamental disadvantage: there are no delay guarantees. As a result, purely contention systems are unsuitable for periodic traffic.

In slotted systems, such as TDMA, all users are effectively synchronized and have different time slots of certain duration assigned to them in periodic fashion. Obviously, this is very well suited for periodic traffic. For bursty traffic, however, channel capacity is wasted, because a station will have a time slot assigned to it even if it does not have data to transmit. A significant problem for TDMA protocols is selecting time slot duration or packet size. If the time slot durations are chosen to match the largest message lengths, shorter messages will not use the channel effectively. On the other hand, if shorter slot sizes are used, the delivery of longer messages will require several time slots. To complete the delivery of longer messages, more time will be required. The message size cannot be known in advance and is likely to change dynamically.

Polling, another type of access method, requires a central station. This central station controls the network by polling individual stations. Stations that have data to transmit will send it in response to a poll. Stations that have periodic data to transmit usually can request to be polled on a periodic basis. The central station typically maintains a global queue of requests. Polling is efficient in the sense that it achieves dynamic resource allocation. However, it has disadvantages as well. First, the overhead for maintaining the global queue can be high. The overhead incurred by the channel access mechanism depends on the number of users, unlike contention schemes. Second, all data must pass through the central station, even if it is not destined for the central station. Third, polling is not suitable for wireless ad-hoc networks, which do not have a central station.

Access protocols can be classified as static or dynamic. The advantage of static allocation MACs is that each user is guaranteed a share of the network resource. The disadvantage is that resources cannot be transferred from one user to another, i.e., users that have no data to transmit are given the same resources as users with data to transmit. MACs with dynamic allocation try to provide network resources only to users that have data to transmit. Compared with other dynamic-allocation protocols, contention-based MACs are easier to implement. This ease comes from the fact that users can join or leave the network at any time. This advantage is the main reason why wired Ethernet is far more popular than the contention-free token ring network. Such a feature is of paramount importance in wireless networks where stations roam about freely. In general, for light-to-medium traffic loads, dynamic-allocation MACs perform better than static-allocation MACs. Under heavy traffic loads, the end-to-end delay that static-allocation MACs achieve is lower than that of dynamic-allocation MACs.

More detailed discussions of multiple-access methods can be found in the references [B9] and [B121].

Wireless communication protocols are affected by the mobility of stations. There are two types of stations to distinguish between—portable and mobile stations. A portable station is one that can be moved from one place to another place, but normally participates in a wireless network when it is in a fixed location. Fixed or slowly moving stations can be considered portable. Mobile stations can participate in a network while in motion.

Thus, the basic requirements of the MAC are:

a) The MAC protocol must be independent of the underlying physical layer.[5]

b) The access mechanism must be efficient for both bursty and periodic traffic.

c) The MAC must handle mobile users.

---

[5] In practical systems, it is not possible for the MAC to be completely independent of the physical layer. The MAC must have some knowledge of the underlying physical environment and certain parameters of the physical layer.

PHY requirements

The PHY is the first layer of the seven-layer OSI model and is responsible for transporting bits between adjacent systems over the air. The PHY performs two functions, depending on whether the device is in transmit or receive mode. In transmit mode, the PHY receives a bit stream from the MAC layer and performs signal processing operations such as error-correcting coding and modulation to convert the bit stream into an electric signal that is supplied to the antenna. In receive mode, the PHY receives the electric signal from the antenna. After signal processing operations such as demodulation and error-correcting decoding, the PHY converts the signal to a bit stream that is passed to the MAC. In addition to these two functions, all IEEE 802 standards require the physical layer to provide carrier-sense indication back to the MAC.

The wireless channel typically leads to bit error rates much higher than for wired channels. Soft-decision decoding provides higher coding gain and is more important for wireless systems with multipath than in additive white Gaussian noise (AWGN) channels. In addition, errors frequently occur in bursts, coinciding with deep fades on the link. Therefore, error-control schemes are important. Three types of error control coding schemes are used:

Error control is a form of diversity, similar to diversity transmission and reception.

 a) Block codes
 b) Convolutional codes
 c) Automatic repeat request (ARQ) schemes

An important point here is that most error-correcting codes are designed to protect against random errors, but not burst errors. A technique that reduces the statistical dependence of errors is interleaving. With interleaving, the symbols contained in one code block are not transmitted in consecutive order, but instead are interspersed among other transmitted symbols so that a signal fade is less likely to impose a dense burst of errors on individual code block. If interleaving can be performed over a sufficiently long time, then errors in individual blocks can be made independent. In practical systems, the latency is limited, and this puts a limit on the interleaver time interval. After decoding to ultimately decide whether there is still an error, cyclic redundancy code (CRC) is performed.

The general requirements of the PHY are bandwidth efficiency and power efficiency. Because of the constant demand for higher data rates, bandwidth efficiency is becoming increasingly important. The importance of power efficiency varies among wireless networks. WLANs are typically used on portable terminals, which are battery-powered only for a limited time. These terminals are normally powered from the AC power sources. Thus, power efficiency is not very important for WLANs. WPANs, on the other hand, are used on mobile devices, which are battery-powered all the time. Power efficiency is of paramount concern. Class-C power amplifiers provide the highest power efficiency. However, they are highly nonlinear and require a constant-envelope modulation scheme. That is why frequency shift keying (FSK) and variations like GFSK are used in some wireless physical layers.

 Power requirements can be divided into two categories: the power needed to operate the electronic circuits and that needed by the power amplifier to supply the antenna with an electric signal of a certain level. The power needed by the power amplifier translates directly into signal coverage. The power needed to operate the electronic circuits depends on the complexity of the signal processing algorithms and the implementation. Only the amount of power needed by the amplifier is usually considered.

PHY layers can be frequency hopping (FH), direct-sequence spread-spectrum (DSSS), or orthogonal frequency division multiplexing (OFDM). Frequency-hopping systems achieve a natural robustness against burst errors. If one carrier is in a fade, the next one is likely not to be in a fade.

Sublayers of PHY and MAC

Both the PHY and the MAC layers can be composed of sublayers. The interface between the MAC and the PHY, like the interface between any two protocol layers, is called service access point (SAP). These service access points are shown in Figure 1–5. A protocol data unit (PDU) is the data unit exchanged between peer entities of the same protocol layer. A service data unit (SDU) is defined as the data unit generated for the next lower layer (in the downward direction) or the data unit received from the previous lower layer (in the upward direction) (Figure 1–5).

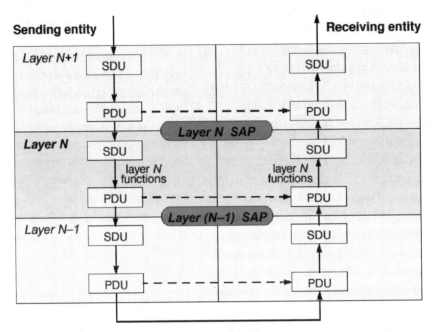

Figure 1-5: PDU and SDU in a protocol stack

The MAC can be considered as providing services to the layer above, and the PHY can be considered as providing services to the MAC layer. In a layered protocol system, the information flow across the boundaries between the layers can be defined in terms of primitives. These primitives represent different items of information and cause actions to take place. These primitives do not appear as such on the air interface; they serve to define more clearly the relations of the different layers. The "request" primitive is an initial request for service from a lower layer. The MAC sends the appropriate dynamic service request message (addition, change, or deletion) across the air link to the receiving MAC sublayer. The receiving MAC sublayer generates an "indicate" primitive to inform the peer convergence sublayer of the request. The peer convergence sublayer responds with a "response" to its MAC, causing it to respond to the initiating side MAC with the appropriate dynamic service response message. Finally, the MAC sublayer sends a "confirm" primitive to the requesting convergence sublayer. At any point along the way, the request may be rejected, terminating the protocol. There are special cases when it is not necessary to send a request over the air link

Chapter 1: Introduction

to the peer station. In these cases the "confirm" primitive is issued directly by the MAC on the originating side. Such cases may occur, for example, when the request does not depend on the other side. The use of these primitives to provide peer communication is shown in Figure 1–6.

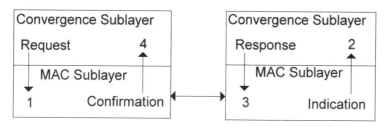

Figure 1–6: Use of primitives to request services

Chapter 2 The IEEE Standard for WLAN: IEEE 802.11

OVERVIEW AND ARCHITECTURE

Work on the first wireless data communications standard began in 1990 when IEEE 802.11 was started as a working group within IEEE 802. The success of Ethernet or IEEE 802.3 and the desire to have "wireless Ethernet" was the motivation for creating IEEE 802.11 as a working group. Wireless Ethernet was originally intended to provide connectivity where wiring was inadequate to support wired LANs. (Ethernet needs special cables called CAT5. CAT5 wiring is not available in older buildings and most homes, nor can it be inexpensively retrofitted.) Another motivation for forming an IEEE 802.11 working group was the allocation of the 2.4 GHz band for unlicensed devices by the FCC in 1986.

The general architecture of IEEE 802.11 is shown in Figure 2–1.

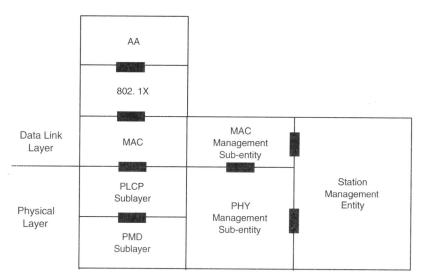

Figure 2–1: General architecture of IEEE 802.11

Figure 2–1 summarizes the architectural view, emphasizing the separation of the system into four major parts: the MAC of the data link layer, the PHY, IEEE 802.1X, and upper-layer authentication protocols. The data link layer consists of an IEEE 802.1X layer and a MAC sublayer. The physical layer consists of two sublayers: a physical layer convergence protocol (PLCP) sublayer and a physical medium-dependent (PMD) sublayer. The PHY and the MAC have management subentities, which communicate with the station management entity.

The IEEE 802.11 working group spent a lot of time on the medium-access protocol of IEEE 802.11 before deciding on carrier-sense multiple access with collision avoidance (CSMA/CA), similar to the CSMA with collision detection (CD) used in IEEE 802.3. The difference between the two is that in wireless communications, collision detection is not possible. A transmitting station cannot reliably detect collisions because the transmitted signal will be considerably stronger than the received signal. The main goal of the MAC is to make the upper layers unaware of the unreliable nature of the wireless channel. This is important because these protocols do not have the concept of mobility. The standard allows a mobile station to roam freely throughout a WLAN and appear to be stationary to the protocols above the MAC.

Recently the MAC of IEEE 802.11 was developed further. First, it was realized that the security mechanism in the original IEEE 802.11 [B68] was totally inadequate. As a response, part of the IEEE 802.11 group, called Task Group I or IEEE 802.11i, began working on an advanced security mechanism. Also, with the advent of multimedia communications requiring QoS, work began on providing QoS support. The corresponding task group is IEEE 802.11e. With the deployment of the first networks, it also became clear that the access points from different vendors must support a common protocol to communicate effectively. The result is IEEE 802.11f, the interaccess point protocol.

As noted previously, six physical layers are currently defined. Two reasons for this number are the constant need for a higher data rate and that IEEE 802.11 works in several different frequency bands. A frequency-hopping PHY and a direct-sequence spread-spectrum PHY, each capable of operation at 1 and 2 Mb/s, were envisioned in the original standard. In addition, an infrared PHY was also standardized. Before even the first standard was completed, the task group realized that the data rates were too low for the technology to succeed in the marketplace. Thus, a new physical layer was created, resulting in IEEE Std 802.11b-1999 The new protocol is able to deliver data rate of 11 Mb/s. The modulation technology in IEEE 802.11b is known as complementary code keying (CCK) modulation. In addition, IEEE 802.11b contains an optional mode, known as packet binary convolutional coding (PBCC). The fifth physical layer, IEEE Std 802.11a, is motivated by the allocation of the UNII bands by the FCC in 1997. Yet another recent work, IEEE 802.11h, adds dynamic frequency selection (DFS) and transmitter power control (TPC) to IEEE 802.11a. The sixth physical layer, completed in 2003, is IEEE Std 802.11g. It provides the same data rate as IEEE 802.11a (up to 54 Mb/s) but works in the 2.4 GHz band. IEEE 802.11g was stimulated by the decision of the FCC to allow digital modulation in the 2.4 GHz band. Thus IEEE 802.11, like any standard in a growing industry, continues to be a work in progress.

The organization of this chapter is as follows. The remainder of this section is devoted to the general architecture of IEEE 802.11. Then the MAC of IEEE 802.11 is presented, followed by a discussion of the physical layers. The MAC presentation is organized as follows. The security mechanism is discussed on page 49, followed by a discussion of the interaccess point protocol on page 66. The section beginning on page 70 is devoted to the QoS mechanism. The section beginning on page 96 is devoted to the physical layers, mainly to IEEE 802.11b, IEEE 802.11a, and IEEE 802.11g. Then the installation of WLANs is discussed on page 117. At the end of the chapter on page 125, current technology and business trends are discussed briefly.

In a way, the goal of the wireless network is to extend the reach of the wired network or backbone.

According to IEEE 802.11, a logical device that participates in a network is referred to as a station. A station consists of a physical layer and a medium access control layer. The basic network is called a basic service set (BSS). One difference with wired LANs is that in wireless networks there are two distinct types of BSSs. The first one is an ad hoc or independent BSS (IBSS). Typical examples of IBSSs are networks formed by personal digital assistants, laptops, CD or DVD players, cell phones, etc. These networks are short lived. The second type of networks, denoted as just BSS, is distinguished by the presence of a special station called access point (AP). The access point allows the network to connect with another network, typically a wired network such as

Ethernet. The backbone network is typically wired, but can also be wireless. Note that the AP is simultaneously a member of two networks—the wireless BSS and the backbone network. In a BSS, client stations communicate only with the access point. Thus, if one station wants to communicate with another station all information must pass through the access point. 802.11 works this way because in most of the applications, client stations rarely communicate among themselves. A typical application is Internet access, where the client stations use the backbone network through the access point. A group of BSSs can be combined to form an extended service set (ESS). Figure 2–2 shows an ESS, consisting of two BSSs. A roaming station in an ESS needs a handoff protocol, which defines how the APs hand off connections for stations.

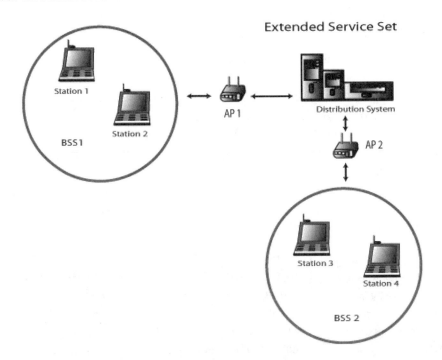

Figure 2–2: Extended service set and distribution system

The standard is designed to support both BSS and IBSS topologies. In a BSS, decision-making is centralized. In an IBSS, decision-making is distributed.

From the architecture of the IEEE 802.11 network it is clear that two groups of functions are called in the standard services. The first group, called station services, are part of every station. The station services are authentication, deauthentication, key distribution, data authentication, replay protection, privacy, and delivery of data. When there is an access point, another group of services, called distribution services, manages traffic. The distribution services are association, disassociation, distribution, integration, and reassociation. Association, disassociation, and distribution control access and provide data confidentiality. Distribution and integration support data delivery between stations.

The MAC layer exchanges three types of messages: data, management, and control. Management messages are used to support the services. Control messages are used to support the delivery of management and data messages.

IEEE 802.11 SECURITY

In this section, the security problem in wireless networks is discussed first, followed by a discussion of solutions offered by IEEE 802.11.

The problem of network security has attracted considerable attention, as networks are increasingly used in business transactions. The security threats that can arise in any network include the following:

a) Unauthorized users can get access to the network.

b) An unauthorized user can get access to the network by convincing the network that it is one of the legitimate users.

c) Information can be received and decoded by eavesdroppers. This information can be later replayed in its original or possibly modified form.

d) An attacker can flood the network with data, resulting in network overload and denial-of-service to authorized users.

e) An unauthorized user can be the man in the middle, convincing part of the network that it is a client (or station), and convincing the other part of the network that it is a server (or an access point).

Before stations can exchange data, a connection must be established. Scan, authentication, and association are necessary to establish a network. Scanning is used to discover existing BSSs that are within range. APs periodically transmit beacon frames that contain timing and network information and can be used to discover BSSs. Immediately after discovering a network, a client cannot join. Before joining a BSS, a client must demonstrate, through authentication, that it is allowed to do so. (In a wired LAN, authentication is replaced by the physical connection to the network. In other words, in a wired LAN, if a station can physically connect to the network, it is authenticated.) All wireless networks require authentication, posing a formidable, but solvable security problem. The security mechanism that IEEE 802.11 uses consists of authentication, key distribution, replay protection, and privacy. The first security mechanism—authentication—is the protocol by which stations prove their identities. However, authentication only is not secure enough, because the wireless medium is still open to an attacker even after authentication. Unless a frame is cryptographically protected by a key tied to the original authentication, it is not possible to determine whether the frame was really sent by the association peer. Encryption and decryption form the privacy mechanism, which will be discussed on page 58. The functions that comprise the security mechanism are discussed next.

Authentication

Historically, IEEE 802.11 has defined three types of authentication: open system, shared key, and upper layer. The first two of these are called link-level authentication and were defined in the first IEEE 802.11 standard of 1997. According to open-system authentication, two messages are exchanged. The first message asserts identity and requests the identity of the other station. The second message returns the result of the authentication—success or failure. Open-system authentication is equivalent to trusting everyone, and therefore does not provide any security. Open-system authentication is sometimes referred to as a null authentication algorithm.

Shared-key authentication intends to do better. The goal of shared-key authentication is to authenticate only those stations that know a shared secret key (Figure 2–3). Shared-key authentication assumes that the shared key is delivered independently of the authentication process. Using encryption and decryption, a challenge text is encrypted at one station, transmitted, and decrypted at the other station. If the result is identical with the original challenge text, the stations do, indeed, have the same key, and authentication succeeds.

Chapter 2: The IEEE Standard for WLAN: IEEE 802.11

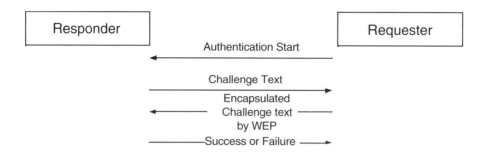

Figure 2–3:   Shared-key authentication

Clearly, before shared-key authentication is performed, stations must have at least one common key. These common keys must be distributed through a separate secure channel. The first disadvantage of shared-key authentication is that it is only one-way authentication. At one time, this was acceptable because access points did not require authentication. This left the BSS open to the presence of rogue access points, which clearly compromised security. The second disadvantage of shared-key authentication is the use of static keys. There is no key management. Furthermore, keys are defined as 40-bit quantities, which even in 1997 was too short, but was the maximum key length permitted for export outside the U.S. and Canada by U.S. government regulations.

Shared-key authentication, however, has an even more serious disadvantage. It fails to achieve even its objective—to authenticate stations that possess the secret key. The reason is that it makes public all the information required to reconstruct the key. The conclusion is that shared-key authentication, much like open-system authentication, does not provide any security. The failure of shared-key authentication undermined considerably the trust in wireless networks security. The amendment to the standard, approved in June 2004, as IEEE Std 802.11i™ [B75], provides adequate security, although it may take some time to regain the public's trust in wireless security. To convey the message that wireless is indeed secure, companies have chosen a marketing name that the public can understand. Wireless networks that implement the novel security algorithms will be called WPA (wireless protected access). In this subsection, the security mechanism developed by IEEE P802.11i will be discussed.

Chapter 2: The IEEE Standard for WLAN: IEEE 802.11

Some background is necessary to understand the development of P802.11i. Because other standards organizations, e.g., the Internet Engineering Task Force (IETF) have worked on advanced security mechanisms, IEEE 802.11 decided to leverage that work. This avoids duplicating functions at the MAC sublayer that are already performed at higher layers. Thus, a robust security network utilizes non-802 protocols for its authentication and key management services. As shown in the IEEE 802.11 architecture in Figure 2–1, these protocols reside above the MAC layer. A robust security network requires three components, which have been developed outside of IEEE 802.11.

The first new component is an IEEE 802.1X Port [B67]. IEEE 802.1X Ports determine when to allow traffic across an IEEE 802.11 wireless link. They reside above the IEEE 802.11 MAC and all data traffic that flows through the MAC also passes through the IEEE 802.1X Port. The architecture of IEEE 802.11 includes an IEEE 802.1X port (Figure 2–1). IEEE 802.1X requires higher-level authentication and provides key management.

Before authenticating any supplicants, the authenticator and AS must authenticate each other. Therefore, it is assumed that a secure channel exists between them. It is important to note that the AS is a logical entity; in real implementations it may be convenient to be integrated into the same physical device as an AP.

The second new component is the Authentication Agent (AA). This component resides on top of the IEEE 802.1X Port (Figure 2–1) at each station and provides for authentication and key management. The Authentication Agent utilizes protocols above both the IEEE 802.1X and IEEE 802.11 layers to provide its services.

The third new component is the Authentication Server (AS). The AS is an entity that resides in the distribution system that participates in the authentication of all stations (including access points) in the Extended Service Set. The AS provides material that every station in the robust security network can use to authenticate every other station. The AS communicates with the AA on each station.

A robust security network, illustrated in Figure 2–4, is a network supporting upper-layer authentication based on IEEE 802.1X. A robust security network has the following characteristics:

Chapter 2: The IEEE Standard for WLAN: IEEE 802.11

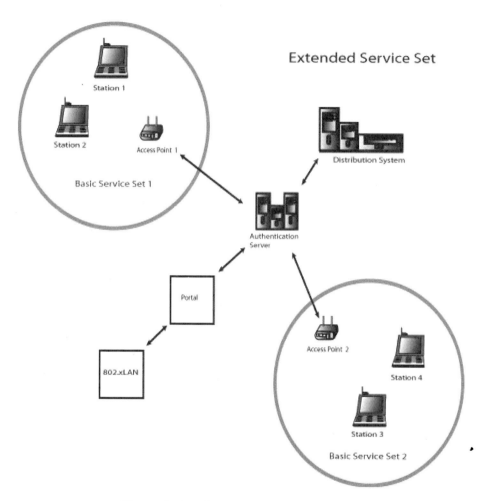

Figure 2–4: A robust security network

a) Mutual authentication between access points and stations.
b) Key management algorithms (Cryptographic keys age and must be refreshed).
c) Cryptographic key establishment.
d) An enhanced data encapsulation mechanism, called Counter-Cipher-block chaining Message authentication code (CCM), and optionally temporary key integrity protocol (TKIP).

First let's consider what is IEEE 802.1X and how it is used with 802.11. IEEE 802.1X "Port-Based Network Authentication" was originally designed for networks, in which undetected eavesdropping is not possible [B67]. However, the standard does not prohibit operation over shared-media networks. As IEEE 802.11 LANs increased in popularity, the need for a proper authentication and key management presented itself, and it was natural to want to leverage mechanisms already been defined in another IEEE 802 standard (802.1X).

IEEE 802.1X is a standard for port-based network access control. A port is a point at which a system attaches to a LAN. Through its ports a system can access services offered by other systems on the LAN, and also it can export services to other systems on the LAN. The system's ports are controlled to ensure that only authorized systems access its services. IEEE 802.1X-2001 is based on the Extensible Authentication Protocol (EAP)[6] over LANs, also known as EAPOL. EAPOL is used to exchange EAP messages. The EAP entities are supplicant, authenticator, and AS. These entities exchange EAP messages to authenticate the supplicant. A port of a system can be either an authenticator or a supplicant. The port configured to enforce authentication before allowing access to services that are accessible via that port adopts the authenticator role. The port configured to access the services offered by the authenticator's system adopts the supplicant role. The AS performs the authentication function necessary to check the credentials of the supplicant on behalf of the authenticator, and indicates whether or not the supplicant is authorized to access the authenticator's services. As can be seen from these descriptions, all three roles are necessary in order to complete an authentication exchange. Authentication is mutual. Any system can—and in a

---

[6] The EAP was originally designed to support authentication over the Point-to-Point Protocol (PPP), and is a product of the Internet Engineering Task Force (IETF).

secure network, must—authenticate any other system before communicating with it. Therefore, any system is capable of adopting the roles of supplicant or authenticator. An authenticator and an AS can be co-located within the same system, allowing that system to perform the authentication function without the need for communication with an external server.

Upper-layer authentication does not occur as a part of IEEE 802.11 authentications, so it does not appear in MAC-sublayer authentication management frames. Instead, IEEE 802.1X implements upper-layer authentication, by encapsulating an authentication scheme outside of IEEE 802.11. The upper-layer authentication mechanism is shown in Figure 2–5.

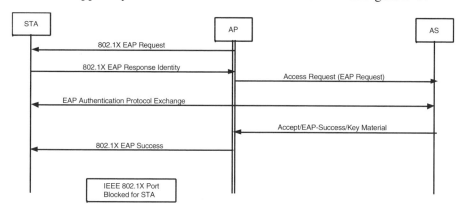

Figure 2–5: EAP Authentication

Because there are three approaches—open, shared key, or upper layer—before authentication, stations must determine which particular approach will be used. In a robust security network, only upper-layer authentication is used, and no authentication operates at the MAC sublayer itself. It should be noted that the choice of an acceptable authentication protocol is an issue for both access points and stations, because the goal of upper-layer authentication is mutual authentication between the access point and the station, not just authentication of the station to an access point. An AP that supports

 Mixing the two authentication methods by performing link-level and upper-layer authentication simultaneously is not allowed. A station's attempt to use IEEE 802.11 authentication when upper-layer authentication has been negotiated should be considered as a service attack.

advanced security advertises its capability by asserting appropriate bits in the security subfield in its beacons and probe response messages. It may also advertise the explicit authentication methods that it supports. In the passive or active scanning process stations detect the capability of the AP and respond accordingly. Typically a client station will choose an authentication method from the list of supported authentication methods by the AP. Because the advanced security mechanism is new, it can only be supported by APs and stations manufactured in the future. Therefore, for some time there will be mixed networks in which some of the stations support robust security measures and some do not.

If an AP does not support advanced security (legacy AP), stations that support advanced security may use the IEEE 802.11 MAC authentication algorithm or may decline to associate. An AP that has robust security capabilities can support legacy stations. Legacy stations ignore the appropriate bits in the Beacons and Probe Responses and do not assert them in their own Association and Reassociation Requests. Similarly, legacy APs do not advertise robust security in their Beacons and Probe, Association, and Reassociation Responses. In this way, new equipment can identify and interwork with legacy equipment. Clear legacy stations can use the legacy MAC sublayer authentication (open or shared key) and then associate. Any station or access point might choose not to associate with another device for many reasons. Networks, in which some equipment supports advanced security algorithms and some does not, are called transition security networks. It must be emphasized that transition security networks are not secure, because compromising the security between devices capable of robust security and legacy devices equipment can compromise even the connections among devices capable of robust security. Only a network that allows robust security network associations is a robust security network

The security mechanisms depend also on the network architecture. In an IBSS, each pair of stations selects the security algorithms to be used, and each station must make its own authentication decision. In an ESS, the AP can enforce a uniform security model. In an ESS, the station initiates all associations. In an IBSS, a station must be prepared for other stations to initiate communications. Thus, a station in an IBSS can negotiate its desired security algorithms when it accepts an association initiated by another station, while in an ESS the AP always chooses the security suite being used. In a secure ESS, the AP may delegate the authentication decision to a physically separate AS, while in an IBSS each station must make its own authentication decision regarding each peer. There is no architectural difference between the two; as in the IBSS case, every station implements its own AS.

It must be noted that IEEE 802.11i does not specify a mandatory authentication method. There are several upper-layer algorithms that can be used. EAPOL is one such protocol. Other protocols can also be used as long as they satisfy two requirements: (1) The authentication algorithm must perform mutual authentication, and (2) it must provide for key agreement based on the authentication. A practical requirement towards any algorithm is to be fast enough not to affect the overall QoS. For example, voice over IEEE 802.11 applications require authentication to be completed within approximately 20 ms.

At present there are several Extensible Authentication algorithms used in commercial products, offered by companies such as Cisco®, 3Com, Lucent®, and others. EAP-Cisco (aka LEAP) is password-based, EAP-TLS (Transport Security Layer) is certificates-based, and EAP-PEAP (Protected EAP) and EAP-TTLS (Tunneled TLS) are hybrid, based on certificates and password. The open architecture of the standard ensures that future algorithms can also be used.

Once authentication completes, IEEE 802.1X allows data traffic beyond IEEE 802.1X traffic and stations can associate. Association creates a logical connection between a station and the AP. Once association is established, the AP will buffer, deliver, or forward traffic to the station. A single IEEE 802.1X Port corresponds to one association, and each association maps to one IEEE 802.1X Port. The IEEE 802.1X Port does not permit general data traffic to pass between the station and the AP until after an authentication procedure completes. Either the station or the AP can use disassociation. A station sends a disassociation message when it is leaving the BSS. An AP may send a disassociation message to a station if it has no

> Authentication can be terminated by deauthentication. To terminate an existing open or shared-key authentication, the deauthentication service within IEEE 802.11 is invoked. To terminate an upper-layer authentication, deauthentication is also an upper-layer function.
>
> In an ESS using open or shared-key authentication, IEEE 802.11 authentication is a prerequisite for association. Hence, the act of deauthentication causes the station to be disassociated.

resources to handle this station. Disassociation makes the AP unreachable to the station, and vice versa. Reassociation is similar to association. A station roaming in an ESS uses reassociation to establish a connection with a new AP, identifying in the message the AP that it was previously associated with.

Note that although, logically, association takes place after authentication, in a network with higher-level security, stations are required to associate without any MAC sublayer authentication. In this way, the upper-layer authentication process is allowed to take place above the MAC layer. Authentication packets (contained in IEEE 802.11 MAC data frames) are passed via the IEEE 802.1X uncontrolled port. However, the association exists only for a period of time sufficient for authentication to take place. If authentication is not completed within that time, the station noticing the delay will disassociate. Nevertheless, this seems a weakness in the standard, because the 802.1X port cannot filter out unauthorized packets during this period of time. With this model, IEEE 802.1X, rather than the IEEE 802.11 MAC, makes decisions as to which packets are permitted onto the Distribution System. If the association or reassociation completes successfully with the selection of upper-layer authentication, the IEEE 802.11 MAC passes all data packets it receives from higher layers, delegating the filtering of any unauthorized traffic to IEEE 802.1X.

APs use the distribution service to forward data from a station in its BSS to another station within the BSS, or the ESS, or to a router for delivery outside the WLAN. Integration takes an IEEE 802.11 frame and recasts it as a frame for a different type of a network, such as Ethernet.

Privacy

Privacy (or encryption) is required to achieve security. IEEE 802.11 provides three cryptographic algorithms: Wired equivalent privacy (WEP), Temporal Key Integrity Protocol (TKIP), and Counter-Mode/CBC-MAC protocol (CCMP).

The cipher suite specified in the original IEEE 802.11 is called WEP. The encapsulation algorithm is illustrated in Figure 2–6. Encapsulation here means the way cryptographic data is constructed from plaintext data; decapsulation is the reverse process. The WEP algorithm is a form of stream cipher, in which plaintext is bit-wise XORed with a pseudorandom key sequence of equal length.

In Basic WEP, the per-packet key (also called a seed) is simply a concatenation of an initialization vector (IV) and a private encryption key. (Figure 2–6.) A new IV is selected for every packet, but the encryption key remains the same. Decryption (not shown) is simply the reverse of the process in Figure 2–6. The same algorithm can be used for keys of different lengths. WEP is only an encryption algorithm, it does not have data integrity keys. Note that shared-key authentication can be implemented only if WEP, which is optional, is implemented.

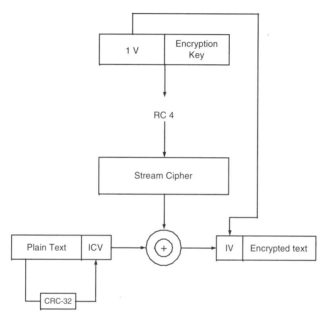

Figure 2–6: WEP Encapsulation algorithm

In basic WEP, not only is the 40-bit key short, but the key remains constant across several frames. Because the same plaintext encrypted with the same key always results in the same ciphertext, it seems possible that the key can be computed. Several researchers investigated this weakness of WEP in 2001 and confirmed that this assumption is correct. Because the number of possible keys is not large enough, a passive attacker can derive the WEP key by only listening to the network and performing statistical analysis. Another approach to derive the key, suitable for an active attack, is to solicit information from the network.

For example, an attacker can send a text to another station and then will listen waiting to see the ciphertext. The key is simply derived from the ciphertext and the original text. Another possible way is to capture a frame, change a few bits in the frame and resend the frame. This will likely result in a higher-layer protocol error. Again, the attacker waits for the error ciphertext to derive the key. The WEP algorithm provides security only against casual monitoring. It offers no protection against cryptographically sophisticated enemies. Note also that the WEP algorithm itself does not provide data integrity guarantees. It is possible to compute an Integrity Check Value (ICV) and embed ICV in the plaintext before encryption.

TKIP and CCMP are new cipher suites, developed by IEEE Std 802.11i (approved in June 2004). WEP and TKIP are based on the RC4 algorithm. CCMP is based on Advanced Encryption Standard (AES). A means is provided for stations to select the algorithm to be used for a given association. In IEEE 802.11 jargon the MIC is also affectionately called "Michael."

Encryption provides a data privacy function only. In particular, it does not afford any protection against data modification. To provide data authenticity requires the use of a data authenticity mechanism. The way to accomplish data authenticity is to compute a tag, called a Message Integrity Code (MIC), using a keyed cryptographic function. Without the MIC, it would be easy for an attacker to intercept a message, flip a few bits in the message, flip the corresponding bits in the ICV to get it to match, regenerate the CRC, and forward the packet on to the ultimate destination. Note that the attacker would not have to know the encryption key to carry out this attack, and the attacker does not need to decrypt the message to carry out the attack. These message-forgery attacks attracted considerable debate in the IEEE 802.11i Task Group. The MIC protects against that. This tag is transported with the data over an unprotected channel with the data it protects, and its value verified by the receiver, using the same key with the same cryptographic function. If the integrity check value is successful, the responder responds with "success." However, even if an attack fails, it still gives some information about the key. It becomes feasible to recover the key after many attacks against the key. Therefore, a MIC failure indicates an attack. After the first MIC failure, stations are required to disassociate and rekey. To limit the probability of a successful attack, any station that detects two MIC failures within 60 seconds must stop all communication for 60 seconds.

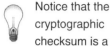
Notice that the cryptographic checksum is a one-way function in the sense that it is extremely difficult to find a message to produce a given checksum.

Stations using CCMP or TKIP can authenticate the data origin and can detect replays. Authenticating the data origin means that when receiving data from another station, the recipient can determine whether the data actually was sent by that station. The replay detection mechanism detects whether the frame is an unauthorized retransmission.

TKIP ensures strong encryption by including (1) dynamic key management, where the key for every packet is different, and (2) a 64-bit message integrity check (MIC).

A 64-bit MIC provides significant protection against message forgery. There are possible MICs and if we assume a message size of 1000 bits, there would be 2^{1000} possible messages. This means that $\frac{2^{1000}}{2^{64}} = 2^{936}$ messages produce the same MIC. This does not mean that it is easy to find a message with the same MIC. One needs to make 2^{64} attempts to find a message that will produce a given MIC. If it takes 1 μs to generate a message and compute its MIC, it will take over 10^{12} years to find a message that will produce a given checksum.

TKIP was designed for hardware devices supporting WEP only. IEEE 802.11 does not recommended using TKIP. TKIP modifies WEP as follows:

a) To defend against forgery attacks, there is a MIC.

b) There is a TKIP sequence counter (TSC). The receiver drops frames received out of order. This is a weak form of replay protection.

c) TKIP uses a cryptographic mixing function to combine the TSC, a temporal key, and TA into the cryptographic key.

The encapsulation and decapsulation mechanisms are shown in Figure 2–7 and Figure 2–8.

IEEE Std 802.11i requires using the Counter-Mode/CBC-MAC protocol (CCMP). It provides all four security services: authentication, confidentiality, integrity, and replay protection. CCM employs the AES encryption algorithm using the CCM mode of operation. The CCM mode combines counter mode (CTR) for confidentiality and Cipher Block Chaining Message Authentication Code (CBC-MAC) for authentication and integrity. (The AES is described in [B113].)

The AES is used with a 128-bit key. CCM is a generic mode that can be used with any block-oriented encryption algorithm. CCM needs two parameters: the size of the MIC M, and the length of the frame. CCMP supplies the values $M = 8$, meaning the MIC is eight bytes, and $L = 2$, meaning that the length of the frame can be represented as a two-byte number. CCM requires a fresh temporal key (TK) for every communication session and a unique number, called a nonce, for every frame. A nonce is a number that is pseudo-randomly generated and does not repeat within a frame. If a nonce repeats within a frame, all security protection is lost.

The CCMP encapsulation process, shown in Figure 2–9, consists of the following steps:

1) The packet number (PN) is incremented for every frame. The PN must never repeat for the same TK.

2) Additional authentication data (AAD) is computed from the MAC header. In this way the CCM algorithm provides integrity protection for the fields included in the AAD. MAC Header fields that may change when retransmitted are excluded when the AAD is computed.

3) The CCM nonce is constructed from the PN, A2, and the priority of the frame.

4) The CCMP header is computed.

5) The ciphertext and the MIC are formed from the TK, AAD, nonce, and plaintext data. CCM encryption needs four inputs: a 13-byte nonce, key, AAD, and frame body. The encapsulated frame is formed by concatenating the original MAC header, the CCMP header, the encrypted data, and the MIC.

Chapter 2: The IEEE Standard for WLAN: IEEE 802.11

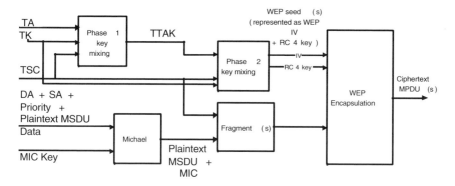

Figure 2–7: TKIP encapsulation diagram

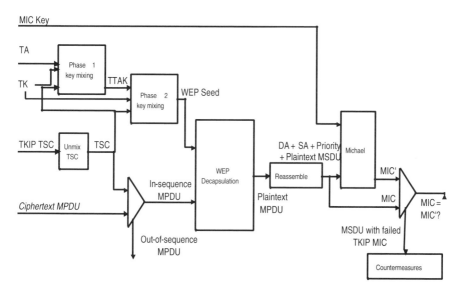

Figure 2–8: TKIP decapsulation diagram

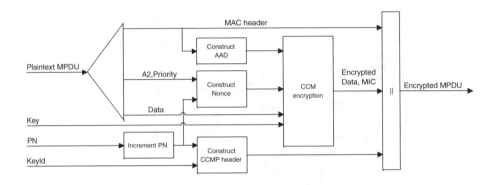

Figure 2–9: CCM encapsulation process

In connection with the traffic classes, introduced by the IEEE P802.11e QoS[7] mechanism, discussed in the following section, the receiver maintains a separate replay counter for each IEEE 802.11 traffic class. The PN recovered from a received frame is used to detect replayed frames. A replayed frame occurs when the PN extracted from a received frame is repeated or not greater than the current traffic class replay counter.

The decapsulation process (Figure 2–10) is based on the following steps:

1) The received encrypted frame is parsed to construct the AAD and nonce.
2) The MIC is extracted for use in the integrity check process.
3) The temporal key (TK), AAD, nonce, MIC, and ciphertext are used to recover the plaintext, as well as check the integrity of the AAD and the plaintext.
4) The receiver extracts the packet number to detect replayed frames. The decryption processing prevents unauthorized replay of frames by validating that the packet number is greater than the packet number maintained for the session. A frame is replayed when the packet number repeats.

---

[7] IEEE P802.11e is not yet an approved standard [B71].

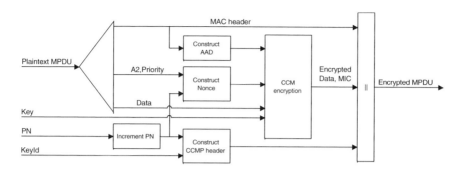

Figure 2–10: CCM decapsulation process

IEEE 802.11 uses the notion of a security association to describe secure operation. A secure association has a finite lifetime. As noted before, the quality of protection with any cryptographic algorithm degrades through key usage. Dynamic key management ensures that the cryptographic keys are fresh. Therefore, a fresh, never-before-used key is required whenever a new "session" begins, so that keys really cannot be used independently of some notion of a session. Fresh cryptographic keys are required for enhanced confidentiality, data authentication, and replay protection. The replay protection counter assumes that peers somehow synchronize a fresh key whenever they reinitialize their replay state, again to avoid having to maintain history of every data frame received.

In an ESS, after a station selects an AP with an appropriate SSID, it uses open-system authentication, followed by association. Negotiation of the security parameters takes place during this stage. Then mutual IEEE 802.1X authentication takes place, followed by key exchange. The AP and the supplicant use two key confirmation handshakes, the four-way handshake and group-key handshake, to indicate that the link has been secured by the keys so is safe to allow normal data traffic. The handshake is performed using EAPOL. In addition to the pairwise keys used for unicast frames, there are group keys used for broadcast messages. Once the keys are known to be established by both association peers, the MAC must discard any data received over the association that is unprotected by the encapsulation algorithm. Without this provision, it is trivial for an attacker to forge a valid message by simply sending a cleartext message.

Chapter 2: The IEEE Standard for WLAN: IEEE 802.11

 As shown in Figure 2–5, the authentication process creates cryptographic keys shared between the supplicant and the AS. A fresh, never-before-used pairwise master key (PMK) is generated and distributed to the AP. The PMK is 256 bits. From the PMK, a Pairwise Transient Key (PTK) is generated. The PTK is partitioned into EAPOL-Key MIC and encryption keys, and temporal keys used by the MAC to protect unicast communication between the authenticator and the supplicant.

Note that roaming must be considered with the security policy. When a station roams in an ESS and uses reassociation, the station must remove the cryptographic key shared with the old AP. A roaming station can cache several keys for several APs, and in this case it will not have to do a complete IEEE 802.1X authorization when it reassociates with each.

The IEEE 802.11 Advanced Encryption Standard Privacy algorithm identifies unicast key "sessions" with IEEE 802.11 associations and uses (re)association messages for the synchronization function, and a random nonce exchange and key derivation to refresh keys. An IBSS desiring to use the Advanced Encryption Standard Privacy algorithm must therefore implement (re)association messages.

The Advanced Encryption Standard Privacy algorithm architecturally lies above the IEEE 802.11 retry function. This is required because data may be accepted by the local IEEE 802.11 implementation, but its acknowledgment may get lost in transit to the peer. If the Advanced Encryption Standard Privacy algorithm were to lie below the IEEE 802.11 MAC retry function, then it would be impossible to recover from this state because the replay protection function would discard all further retries.

INTERACCESS POINT PROTOCOL (IAPP)

Confining stations to a single BSS has little benefits. It is much better to allow stations to roam within an Extended Service Set (ESS), as illustrated in Figure 2–11.

To achieve roaming capabilities, the APs must communicate among themselves. The APs are connected via an abstract layer, called distribution system (DS). The protocol that is used across the DS is new and is called interaccess point protocol.

This protocol, defined in IEEE Std 802.11F, which was approved in June 2003, is the subject of this section.

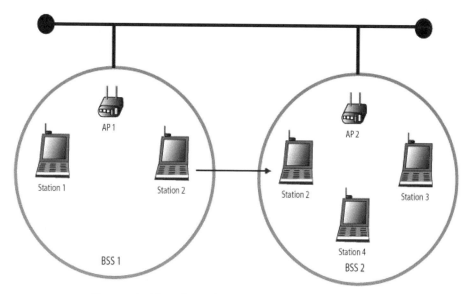

Figure 2–11:   A station roaming within an ESS

Initially a protocol for AP-to-AP communication was not defined by IEEE 802.11. While leaving this protocol to vendors allows great flexibility in DS and AP functional design, APs from different vendors may not work together. In particular, the requirement that a mobile station has a single association at any given time is unlikely to be enforced. As IEEE 802.11 systems have grown in popularity, this limitation has become an impediment to WLAN market growth.

The IAPP affects not only APs. It also affects other devices in the network such as layer 2 networking devices, (e.g., bridges[8] and switches). From a business perspective, this gives opportunities for vendors to differentiate themselves in the market.

---

[8] Large local-area networks are often divided into segments. Bridges are devices used to connect these segments. Access points as members of an Ethernet network could be connected by bridges. Bridges are similar to repeaters. However, repeaters work at the physical layer, while bridges work at the MAC layer. The devices that perform a similar function at the network layer are called routers.

The first goal of IAPP is to ensure that a station is always associated with, at most, one AP, as it roams within the ESS. The goal of single association is achieved using a registration protocol developed by the IETF, called Service Location Protocol (SLP). The registration service can be hosted on an independent server or can be contained within an AP. The second goal of IAPP is to make the APs interoperable. The packets between APs are carried over using the user datagram protocol over IP (UDP/IP). Another objective of the IAPP is not to make IAPP a routing protocol. In other words, the IAPP should not directly deliver the data frames to the stations. Instead, the DS uses existing network functionality to deliver data frames. Therefore, the IAPP requires IEEE 802.11 stations to maintain a network layer address, e.g., IP address. When stations maintain a valid network layer address, with respect to forwarding traffic to stations, APs function much the same as IEEE 802.1D bridges. Additionally, the IAPP maps the address of APs (their BSSID) to DS network layer addresses, i.e., to their IP addresses.

The main operational program of the AP is called AP management entity (APME). The IAPP communicates with the local AP management entity. Several IAPP service primitives are defined that allow the AP management entity to cause the IAPP to perform some function or to communicate with other APs in the DS or a registration service. Several other service primitives indicate to the AP management entity that operations have taken place at other APs in the DS that can have an effect on information local to the AP. The AP management entity must be able to find and use a registration service. It must register as part of an ESS and look up the IP addresses of other APs in the ESS. The AP management entity uses the Service Location Protocol to accomplish this.

The IAPP provides services to an AP in which it resides through the IAPP Service Access Point (SAP). The SAP allows the management entity of the AP to invoke IAPP services and receive indications of service invocations at other APs in a single ESS. Recall that Figure 1–6 shows the four service types that exist at the SAP—requests, confirms, indications, and responses. Service requests and responses are issued to the IAPP by the entity at the next higher layer. The next higher layer is understood to be APME. Service confirms and indications are issued by the IAPP to the AP management entity.

Suppose a station is roaming in an ESS, as in Figure 2–11. Association with a new AP can be accomplished by two commands – association or reassociation.

If association is used, the address of the old AP is not specified and remains unknown by the IAPP. If reassociation is used, the station can inform the new AP of the address of the old AP, because the reassociation command contains the address of the AP the station was previously associated with. The IAPP works differently, depending on whether the roaming station chooses to use association or reassociation. Every time the station associates with a new AP, the APME generates an appropriate request to the IAPP. In turn, the IAPP attempts to enforce the single-station association requirement by sending an IAPP packet called ADD-notify to the DS. This packet notifies other APs and layer 2 devices (bridges) about the association.

This notification causes two changes. First, the forwarding tables will update. All future traffic for the newly associated station is forwarded to its new AP. Second, every AP that receives the ADD-notify packet checks whether the roaming station was not previously associated with it. If this is the case, the AP disassociates with this station and removes any context stored for the station.

A different request is issued when a station reassociates with an AP, since in this case the AP with which the station was previously associated is known. Again, forwarding tables are updated and the DS is notified of the reassociation. This time, however, the AP notifies the IAPP of the BSSID of the AP with which the station was previously associated. The IAPP translates the BSSID to the IP address of the AP. Using the IAPP protocol, a packet called MOVE-notify is sent over UDP and IP. This packet is sent from the AP directly to the old AP with which the reassociating mobile station was previously associated. This guarantees that the old AP will be notified of the new association.

All information stored at the old AP is forwarded to the new AP. The IAPP relies on a station making use of the Reassociation Request when roaming from one AP to another, instead of an Association Request. The reason is that with an Association Request, the IAPP does not specifically notify the old AP of the new association. This may result in the old AP maintaining context for the station that has roamed to a new AP for a longer time than is necessary. This maintenance not only wastes resources at the old AP, but, more importantly, it violates the single-AP association requirement of IEEE 802.11. The reason the old AP may not be notified of the new association is that the ADD-notify packet is sent, using the IAPP protocol over UDP and IP, on the local LAN segment. The packet is sent to the subnet broadcast address, so that it will reach every device on the DSM local subnet. If the old AP is not on the local subnet, it will not receive this packet.

Note the security implications of roaming. When a station reassociates with a new AP, it must first authenticate with the new AP. If authentication with the new AP fails, the station is still authenticated and associated with the old AP. The old AP will not be informed of the unsuccessful authorization attempt. It is possible for authentication with the new AP to succeed, but reassociation to fail. In this case, the network will consider the station associated with the old AP.

While the IAPP is intended to transfer context among IEEE 802.11 access points, it can also be used to transfer context between access points supporting other standards, such as within the IEEE 802.15 and IEEE 802.16 families of standards. Furthermore, it can be used to support the transfer of authentication, authorization, and accounting (AAA) context between devices supporting IEEE 802.1X authentication. The IAPP is independent of the security scheme defined in IEEE 802.11. Authentication information is not carried between the APs. The message exchanges of the IAPP are not protected nor authenticated beyond the protection and authentication afforded by other protocol traffic in the ESS, unless those services are provided by the protocol carrying the IAPP messages. Usually, using IPSec to carry the IAPP messages would provide this protection.

IAPP, however, supports mobility among APs within a single ESS. Higher-layer handoff is required to support mobility among multiple ESSs.

MEDIUM-ACCESS MECHANISM AND REAL-TIME TRAFFIC OVER IEEE 802.11

When IEEE 802.11 started work on the MAC in 1990, it was envisioned that it would be used for computer data communications primarily; therefore, real-time data such as voice and audio was not given much consideration. Because of its CSMA nature, the MAC is optimal for computer data communications, but not for real-time data. Recently, however, data communications systems have been used increasingly to transport real-time data such as voice, audio, and video. All telecommunications service providers are looking at ways to generate additional revenue from their communication networks. Providing different levels of quality of service (QoS) is widely recognized as one of the most important ways to achieve this. The nature of real-time traffic is different from data applications. The main QoS factors are:

- Bandwidth
- End-to-end delay or latency
- Latency jitter
- Signal quality and packet loss

QoS is also a business issue, and the business requirements are:

- Capability to offer service differentiation
- Service availability.

Real-time data has strict throughput, delay, and jitter requirements. For example, to maintain acceptable quality of a human conversation the latency for voice packets must be less than approximately 250 ms or 200 ms, and must remain constant. If a packet arrives with a delay exceeding this amount, it can be discarded, because it loses its value. Latency is usually addressed first by signal processing and protocol improvements. Jitter can be removed by buffering, though a large buffer may create latency problems. Networks that satisfy these requirements are said to provide QoS. Real-time data introduces different requirements towards the network.

Not all of these requirements are stricter. Real-time traffic relaxes certain requirements imposed by data traffic. In particular, real-world signals such as audio and video are somewhat tolerant of bit errors. Voice quality and video quality, as perceived by the human auditory and visual systems, are not severely affected by occasional bit errors. Therefore, real-time data tolerates much higher Bit Error Rate (BER) than asynchronous data. Data packets, of course, do not tolerate any bit errors.

 Note that at present networks, including IEEE 802.11, do not take advantage of this property. In particular, if a packet has a few bit errors, it will be discarded and not passed on to the higher layers, even though the voice or video signal quality might be acceptable to a human recipient.

The QoS mechanism for WLANs is very different from the QoS mechanism in other networks in several respects. First, QoS is an objective, not a guarantee. It may be impossible, or may become impossible, for the requested service quality to be provided. Second, due to the characteristics of the wireless medium, broadcast and multicast data may experience a lower quality of service, especially with regard to loss rate, compared to that of unicast data frames. Third, it becomes necessary for higher layer protocols to become "WLAN aware," at least to the extent of understanding that the QoS characteristics of a WLAN are subject to frequent, and sometimes substantial, dynamic changes. This seems to violate a fundamental objective of IEEE 802.11, but nevertheless is mandated by the wireless environment.

In this section, the different timing intervals that play a role are defined. Then the basic access mechanism for IEEE 802.11 is described. After this, the recently defined QoS facility is described.

For the medium-access protocol timing is very important. The time interval between frames is called interframe space (IFS). There are five timing intervals. From the shortest to the longest they are: short IFS (SIFS), point coordination function IFS (PIFS), distributed IFS (DIFS), arbitration IFS (AIFS), and extended IFS (EIFS).

SIFS and the slot time are fixed by the PHY. These timing intervals for the three main physical layers of 802.11 are given in Table 2-4. One reason for having SIFS is that at any given time, a device can only transmit or receive, but cannot do both. Furthermore, it takes some time for a device to switch from receive to transmit mode and back. This amount of time is called turnaround time. SIFS is equal to the sum of the turnaround time of the RF transceiver, the MAC processing delay, the PHY processing delay, and the RF delay.

PIFS is SIFS plus one slot time. The slot time is the sum of the time for the clear channel assessment, the turnaround time, the MAC processing delay and the air propagation time.

DIFS is SIFS plus two slot times. The DIFS is the time that separates the start of the contention window from the end of the previous transmission.

The EIFS is the sum of SIFS, DIFS, and the length of time it takes to transmit an ACK Control frame at the PHY's lowest mandatory data rate. The EIFS is long enough to protect the ACK from colliding with a transmission from a station that was unable to update its network allocation vector (NAV).

Figure 2–12 illustrates the relation between the SIFS, PIFS, and DIFS as they are measured on the medium. Arbitration interframe space (AIFS) is defined only for the QoS facility. The different interframe spaces are independent of the bit rate. It is convenient to think of these time periods as time gaps on the medium. They are fixed for each physical layer.

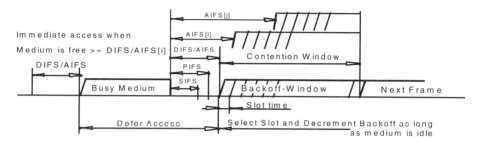

Figure 2–12: The basic access mechanism and interframe space relationships

Central to the medium-access mechanism is the concept of a coordination function (CF). The coordination function is a logical function that determines when a station operating within a Basic Service Set is permitted to transmit and receive data via the wireless medium. The interval of time when a particular station has the right to initiate transmissions onto the wireless medium is called transmission opportunity. The basic access mechanism in IEEE 802.11 is carrier sense multiple access with collision avoidance (CSMA/CA) with exponential backoff. The CSMA/CA mechanism is implemented by the DCF. In this type of mechanism, before transmitting data, a station checks whether the medium is busy, as illustrated in Figure 2–12. If the medium is busy, the station does not transmit. The medium should be available for an amount of time equal to the DIFS if the last frame detected on the medium was received correctly or to the EIFS, if the last frame detected on the medium was not received correctly. After this idle time, every station will generate a random backoff period for an additional deferral time before transmitting. This process minimizes collisions. This is the physical carrier sense mechanism.

The random backoff time is distributed according to a uniform distribution. The maximum extent of the uniform range is called contention window (CW). The distribution is uniform between 0 and CW. The time unit of the backoff timer is called the slot time, which is equal to the maximum round-trip propagation delay. At each time slot, the channel is sensed to determine if there is activity on the channel. If the channel is idle for the duration of the slot, the timer is decremented by one time slot. If the channel is busy, the timer is not decremented. When the channel becomes idle again for a period greater than the DIFS, the backoff procedure continues from where it was interrupted. The process is repeated until the timer reaches zero. Only then will the station begin transmission. After a transmission, a station will wait for an acknowledgment (ACK) back to the transmitter, if a packet is received correctly. The acknowledgment is transmitted after the short IFS. This is necessary to ensure that an acknowledgment is transmitted before a new packet. No station will sense the channel before an acknowledgment is transmitted.

If no acknowledgment is received, the transmitting station will infer that an error has happened. Many circumstances may cause an error to occur—collision, other interference, etc. Error recovery is handled by retransmissions. The transmitting station increments a retry counter and attempts to retransmit. In this case, the contention window is doubled, a new backoff interval is selected, and the backoff countdown begins again. The backoff is exponential to avoid the zero-throughput state where the channel is jammed up with retransmissions. Every station maintains two retry counts: short retry count and long retry count. Their initial value is zero. The short retry count is incremented every time transmission of a short MAC frame (frame of length less than or equal to a certain threshold) fails. This short retry count is reset when a transmission of a short MAC frame succeeds. This long retry count is reset when a transmission of long MAC frame succeeds. The station long retry counter is reset to 0 whenever an acknowledgment frame is received in response to data transmission of length greater than a certain threshold, or whenever a frame with a group address is transmitted.

The contention window is increased every time an unsuccessful attempt to transmit causes either station retry counter to increment, until the contention window reaches a specified maximum. Once it reaches this maximum value the contention window remains at this value until it is reset. This improves the stability of the access protocol under high-load conditions. Retransmitted frames are identified with the Retry field set to 1. Retransmissions continue until the transmission is successful or until the relevant retry limit is reached, whichever occurs first. When the retry counters reach their maximal values, retry attempts cease, and the data is discarded.

Figure 2–13 shows the backoff procedure when several stations have data to transmit. Initially the medium is occupied by station A, and stations B, C, and D defer. DIFS after station A completes its transmission, stations B, C, and D will enter the backoff procedure. The contention is won by station C, which chooses the shortest backoff interval. During the transmission of station C, stations B and D suspend their backoff counters. After station C completes its transmission, B and D again resume decrementing their counters. In the meantime station E also has data to transmit and will enter the contention. This time station D wins the contention. The process continues as shown in Figure 2–13.

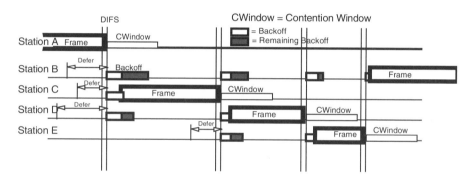

Figure 2–13:   Backoff procedure

There is one situation when a station is not required to perform the random backoff before starting data transmission. A data packet arriving from the higher layer may be transmitted immediately, if the last post-backoff has been finished already (i.e., the queue was empty) and the channel has been idle for a minimum duration of DIFS. All the following packets must be transmitted after random backoff, until the transmission queue is empty again.

However, wireless networks can suffer from a hidden user phenomenon. One station may be able to communicate with two other stations, but these two stations may not be able to communicate between themselves. In this case, a station may sense the channel to be idle, even when there is a transmission. This increases the likelihood of a collision. The physical carrier sense of the CSMA/CA procedure is inadequate to deal with the hidden node problem. However, there is also a virtual carrier sense. The virtual carrier sense mechanism deals with this problem by a reservation mechanism to announce the impending use of the channel. Every frame contains timing values, which are used to maintain special counters, called network allocation vectors (NAVs), which are present in all stations. NAV is implemented as a timer, which is continuously decremented regardless of whether the channel is busy or idle. The NAV is updated from the duration values of all received MAC headers, not just frames addressed to this particular station. Before sensing the medium with the physical carrier sense mechanism, a station would check its NAV. The station would attempt transmission only when the NAV has a value of zero, meaning that there is no transmission on the air. If either the physical or the virtual carrier sense mechanisms indicates that the medium is busy, stations select a random backoff interval.

To help deal with the hidden node problem there is an optional mechanism, called the RTS-CTS mechanism. The transmitting and receiving station can exchange messages called request-to-send (RTS) and clear-to-send (CTS). A station that has data to transmit sends a RTS packet. The intended recipient of the data sends a CTS packet. Both the RTS and CTS packets contain the source and destination addresses as well as the estimated time interval to complete the transmission and to return an acknowledgment.

A station that is not able to communicate with one of the two stations will receive only one packet—the RTS or the CTS. This is enough, however, for it to understand that the channel will be in a busy state and the duration of this state. The duration will be stored in the NAV. It is obvious that the RTS/CTS virtual carrier sense mechanism imposes overhead, which could be significant. The RTS/CTS mechanism is not used for short packets, for which the collision likelihood and cost in terms of retransmission time are small. To determine how short a packet is, there is a threshold called the RTS threshold. If the packet length is below the RTS threshold, RTS/CTS is avoided. In practice, the effectiveness of this reservation mechanism depends on the extent to which stations have identical operating ranges. Note that the RTS/CTS mechanism does not completely eliminate collisions. There may be collisions during the RTS or during the CTS packet, depending on whether the hidden nodes receive the RTS or the CTS packet. Collisions are more likely to occur during the transmission of the RTS signal. Note also that this reservation mechanism is unsuitable for broadcast or multicast packets. If the RTS transmission fails, the short retry count and the station short retry count are incremented until the number of attempts reaches a certain limit.

Note that according to the DCF described previously, all stations are equal. The point coordination function (PCF) is a different coordination function, according to which stations and access points have different roles. While DCF is mandatory to implement, using the PCF is optional. The PCF introduces the concept of a superframe. Superframe is a frame that has two periods: contention-free period (CFP) and contention period (CP) (Figure 2–14). DCF is active in the contention period. PCF is active in the contention-free period. At the beginning of the superframe, if the channel is free, the PCF takes control over the channel. If the channel is busy (and therefore controlled by the DCF), the PCF defers. The PCF operates as follows. Stations can request that the AP registers them on a polling list. Then the AP regularly polls the stations one at a time for traffic. During the CFP, medium usage is entirely controlled by the AP, thereby eliminating the need for individual stations to contend for the medium. However, the AP must contend for the medium, and the start of the CFP can be delayed from its start time. Once the PCF gains access to the medium, the AP sends a beacon frame, begins to deliver queued data to the stations, and can poll stations by sending CF-Poll frames to those stations that have requested contention-free service. Stations that have data to transmit are allowed to send one frame in response to a CF-Poll. Stations that do not have data to transmit simply do not respond to the poll.

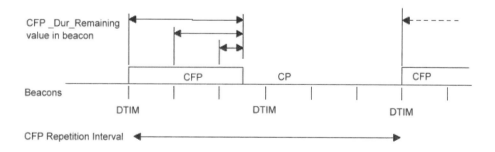

Figure 2–14: Beacons, contention, and contention-free periods

The beacon frame is a management frame that maintains the synchronization of the local timers in the stations and delivers protocol-related parameters. The point coordinator (PC), which is typically co-located with the AP, generates beacon frames at regular intervals. Thus every station knows when the next beacon frame will arrive. The intended period of the beacon frames is called targeted beacon transmission time (TBTT) and is announced in every beacon frame. Note that the beacon frame is required in pure DCF, even if there is only contending traffic. At the beacon period, the AP will schedule a beacon for transmission. If the medium is busy, the beacon transmission will be delayed, according to the CSMA mechanism. When the medium becomes available, the beacon will be transmitted before other data frames (Figure 2–15).

During the CFP, the maximum idle period is PIFS. If the AP does not receive data from a station being polled within PIFS of the polling frame, it can proceed and poll another station.

The PCF can transmit four types of frames: CF-Poll, CF-Ack-Poll, CF-Data-Poll, and CF-Data-Ack-Poll. To increase the aggregate throughput, the polling request is piggybacked on data and/or acknowledgment frames.

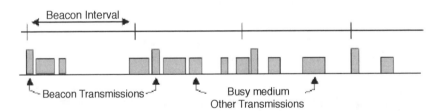

Figure 2–15:   The beacon transmission is almost periodic

IEEE Std 802.11 does not address the question of order when there several stations to be polled, thus providing the opportunity for proprietary algorithms to be used. One approach could be to implement some type of prioritization, where some stations are polled before others. This prioritization can be dynamic. Another approach is to poll in a round-robin fashion, where for example the station with the lowest address is polled first, and so on. The PCF can be considered as a packet-switched, connection-oriented service.

To implement polling the AP must maintain a polling list. If the AP has completed polling all stations and traffic is light the CFP can be ended by the transmission of a CF-end frame. However, there is no mechanism to turn off retransmissions in the CFP.

The beacon frame notifies all of the other stations in the cell not to initiate transmissions for the length of the contention-free period. The length of the CFP can cover multiple beacon periods. Having silenced all other stations, the PCF station allows a given station to have a contention-free access. This allows stations to be even in power-save mode. In an ESS, a station in power-save mode initiates a frame exchange sequence by transmitting a PS-Poll frame to request data from an AP. In the event that neither an acknowledgment frame nor a Data frame is received from the AP in response to a PS-Poll frame, then the station can retry by transmitting another PS-Poll frame at its convenience.

If the AP sends a Data frame in response to a PS-Poll frame, but fails to receive the acknowledgment frame in response to this Data frame, the next PS-Poll frame from the same station may cause a retransmission of the last data. This duplicate data is filtered at the receiving station using the normal duplicate frame filtering mechanism. If the AP responds to a PS-Poll by transmitting an acknowledgment frame, then responsibility for the Data frame delivery error recovery shifts to the AP. The data is transferred in a subsequent frame exchange sequence, which is initiated by the AP. The AP will deliver one block of data to the station that transmitted the PS-Poll. If the power-save station that transmitted the PS-Poll returns to power-save mode after transmitting the acknowledgment frame in response to successful receipt of this block of data, but the AP fails to receive this acknowledgment, the AP will retry transmission of this block of data until the relevant retry limit is reached.

If a hidden station misses the previous beacon frames, then this hidden station will not have any knowledge about the TBTT and will not stop its operation based on DCF. If it has any data to transmit, the result will be collision during the CFP. The 802.11 standard tries to remedy this to some extent by requiring a station to set its NAV at the TBTT, irrespective of the reception of a beacon frame. However, if a station has not received any of the previous beacon frames, it cannot set its NAV at TBTT.

Beacons are transmitted even in an IBSS, in the absence of an AP. Because an AP is not present, there isn't an active point coordinator. However, every station at the TBTT selects a random delay and queues a beacon frame for transmission. When a beacon is received the queued frame is canceled. This scheme guarantees that every station in an IBSS will either receive a beacon or transmit one.

The PCF and DCF operate simultaneously and the PCF has priority over the DCF. The PIFS is shorter than the DIFS, so that the PCF will seize the medium before the DCF does.

The DCF and the PCF are the methods for medium access in the original IEEE 802.11 standard. The PCF was intended to carry periodic traffic such as speech, audio, and video signals. However, this scheme does not provide sufficient QoS to carry real-time traffic such as voice and video for several reasons.

Three main problems with the PCF led to the current activities within IEEE 802.11e to enhance the protocol: unpredictable beacon delays, unknown transmission duration of the polled stations, and lack of prioritization. The contention period may be of variable length and in turn, this may cause the contention-free period to start at different times. At TBTT, a point coordinator schedules the beacon as the next frame to be transmitted, and the beacon can be transmitted when the medium has been determined to be idle for at least PIFS. Depending on the wireless medium at this point in time—i.e., whether it is idle or busy around the TBTT—a delay of the beacon frame may occur. The time the beacon frame is delayed (i.e., the duration it is sent after the TBTT) postpones the transmission of time-bounded data that has to be delivered in CFP. This makes it impossible to maintain periodicity.

Another problem with the PCF is the unknown transmission time of polled stations. A station that has been polled by the PC is allowed to send a single frame that may be fragmented and of arbitrary length, up to the maximum of 2304 bytes (2312 bytes with encryption). Also, because different modulation and coding schemes are specified in the physical layers, the point coordinator does not control the duration of the transmission after polling. This destroys any attempt to provide QoS to other stations that are polled during the rest of the CFP.

Another disadvantage of the PCF is that there is no prioritization of data. If one station has several types of data to send, there is no distinction among them.

The original IEEE 802.11 of 1997 [B68] did not provide QoS when it was approved, and this was expected to be a significant disadvantage for it to succeed in the marketplace. Several influential companies were ready to mass-produce products based on different standards. It turned out that these concerns were unfounded. IEEE 802.11 became the most successful wireless standard, even before the QoS mechanism was in place. One reason the standard is so successful is that data is the primary application. Another reason is that it is possible to improve the original IEEE 802.11 and provide better QoS.

Enhancements in IEEE 802.11e

IEEE 802.11e will bring many enhancements—some general, some specific.

The first general enhancement is an option to allow stations to talk directly to other stations, bypassing the AP, even when there is one. The original IEEE 802.11 standard permits transmitting from one station to another only in ad-hoc mode. According to the original IEEE 802.11, in a BSS, when there is an AP, two stations cannot exchange information directly. If a station A wants to talk to another station B, station A first must send the data to the AP, and then the AP must send it to station B. This is not only inefficient in terms of channel usage, but it doubles the latency. The recent enhancement includes the ability for station-to-station transmission to bypass the AP and move instead directly between the relevant stations within the same BSS. This data transfer is set up using Direct Link Protocol (DLP). This protocol is necessary for several reasons. First, the protocol is used to exchange data rate and security information. DLP is also necessary because the intended recipient may be in power-save mode, in which case it can only be woken up by the AP. This protocol prohibits the stations going into power-save mode for the active duration of the Direct Stream. DLP does not apply in an IBSS, where frames are always sent directly from one station to another. The DLP handshake is illustrated in Figure 2–16.

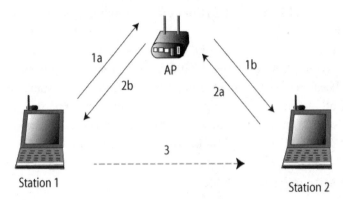

Figure 2–16: Steps involved in the Direct Link handshake.

First, a station 1 that has data to send will use the DLP and will send a DLP-request frame to the AP (arrow 1a in Figure 2–16). This request contains the rate set, and the capabilities of station 1, as well as the MAC addresses of stations 1 and 2. If station 2 is associated in the BSS, the AP will forward the DLP-request to the recipient, (1b). If station 2 accepts the request, it will send a DLP-response to the AP (2a), which contains information about its capabilities. The AP will forward the DLP-response to the first station (2b), after which the direct link becomes active and frames can be sent between the two stations. Note that a protection mechanism such as the RTS/CTS exchange is recommended during the Direct Link handshake.

The second general enhancement is the introduction of negotiable acknowledgments. The existing IEEE 802.11 standard relies on a positive acknowledgment mechanism for all frame exchanges. IEEE 802.11e introduces the notion of negotiable acknowledgments whereby a packet in a stream need not be acknowledged, or such acknowledgments can be aggregated, depending on the parameterization of a stream. Such negotiable acknowledgments lead to a more efficient utilization for the available channel and enable such applications a reliable multicast.

According to 802.11e, there are four acknowledgment types: normal (as in the legacy 802.11), no acknowledgment, no explicit acknowledgment, and block acknowledgment. Clearly, no acknowledgment is used whenever acknowledgment is not required. This could be appropriate for delay-sensitive traffic. Because no acknowledgment is associated with lower reliability of data transfer, it is recommended to use RTS/CTS to increase the likelihood of successful data transfer. The option of no explicit acknowledgment allows acknowledgments to be piggybacked onto different frames.

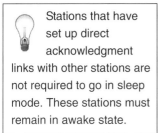
Stations that have set up direct acknowledgment links with other stations are not required to go in sleep mode. These stations must remain in awake state.

The block acknowledgment mechanism allows a block of QoS frames to be transmitted, each separated by a SIFS period. This improves latency and jitter. There are two types of block acknowledgments: immediate and delayed. Immediate block acknowledgment is suitable for high-bandwidth, low-latency traffic, while the delayed block acknowledgment is suitable for applications that tolerate moderate latency.

The third general enhancement is the introduction and traffic prioritization and traffic parameterization. These are central to any QoS mechanism.

Traffic prioritization gives the ability to handle traffic differently according to the priority of that traffic. IEEE 802.11e is using a standard called 802.1D-1998 (specifically Annex H.2 of that standard), which adds three bits to the packet header and gives the LAN the capability to recognize eight levels of priority. The eight priority levels are designated by the integers: 7–0. The priority value can be indicated directly, or indirectly, in a traffic specification. By default, priority 7 is the highest priority and priority 1 is the lowest priority, with priority 0, which is used for best-effort traffic, ordered between priority 3 and priority 2. The resulting default ordering is {7, 6, 5, 4, 3, 0, 2, 1}. Every priority level may have different traffic specifications. A device can implement fewer than eight physical queues. In this case, it must provide a mapping from traffic categories and delivery priorities to the available queues. At least four physical queues are mandatory. These priority values can be specified directly according to Table 2–1, or indirectly according to traffic specification (TSPEC).

Another newly defined ability is the opportunity to handle traffic in accordance with a parameterized specification for each traffic type, allowing for more broadly differentiated traffic than would be available with simple prioritization. Again, the parameterization must be specified as part of the traffic and is not determined by the MAC.

Table 2–1: IEEE 802.1D priority levels

Priority	Traffic type
1 (lowest)	Background (BK) – aperiodic data
2	—
0	Best effort (BE) – aperiodic data
3	Excellent effort (EE) – isochronous data
4	Controlled load
5	Video (VI) – periodic
6	Voice (VO) – periodic
7 (highest)	Network control (NC)

The major novelty developed by IEEE 802.11e is the new coordination function, called hybrid-coordination function (HCF). It is a coordination function that combines aspects of the distributed coordination function and the point coordination function. The hybrid coordinator (HC) updates the NAV and appears to legacy stations as a point coordinator. However, it is a type of a point coordinator different from the one defined in the point-coordination function (PCF). The hybrid-coordination function has a contention-based and a contention-free-based channel access mechanism. It is important to realize that these are not two separate coordination functions; they are part of the hybrid-coordination function. The hybrid coordinator performs bandwidth management including the allocation of transmission opportunities to stations and the initiation of controlled contention intervals. The HCF is called hybrid because it is used during both the contention and contention-free periods.

In IEEE 802.11e, transmission opportunities are defined by a starting time and a maximum duration. These transmission opportunities are won in contention with other stations or granted during the CFP. The QoS CF-Poll from the HC can be sent after a PIFS idle period without any backoff.

The hybrid-coordination function uses a contention-based channel access method, called the enhanced distributed channel access (EDCA). Because it operates in the contention period, EDCF is similar to DCF. OoS support is realized with the introduction of the Traffic Categories (TCs). The prioritized channel access is realized with the QoS parameters per TC, which include AIFS[TC], CWmin[TC], and CWmax[TC]. According to the EDCA, service differentiation is achieved through varying the amount of time a station senses the channel to be idle before backoff or transmission, varying the length of the contention window, and/or varying the duration a station may transmit after it acquires channel access.

The EDCA mechanism provides differentiated access to the wireless medium for the eight delivery priorities. A station desiring to initiate a transfer using the EDCA uses an equivalent to the DCF random backoff time mechanism, except that the station calculates and maintains a backoff timer and contention window for each queue—when there is data to be transmitted for that queue. Before each transmission (when the medium is busy), a station will defer until the medium is determined to be idle without interruption for a certain period of time. This interval is equal to the arbitration IFS for that queue (AIFS[i]) (rather than DIFS) when the last frame detected on the medium was received correctly. Note that AIFS[i] is variable, assigned either by a management entity, or by the AP and measured in time is equal to the integer AIFSN[i] multiplied by the slot time plus SIFS:

$$AIFS[AC] = AIFSN[AC] \times \text{slot time} + SIFS \qquad \text{Eq. 2-1}$$

The integer AIFSN must be greater than 2 for stations and greater than 1 for the AP. In this way, the AP has a higher priority for this channel access mechanism. The values of AIFSN for the different access categories is advertised in the beacons and probe response frames transmitted by the AP. An example is given in Table 2–2.

Table 2–2: Contention window and AIFSN parameters

AC	CWmin	CWmax	AIFSN
BK	CWmin	CWmax	7
BE	CWmin	CWmax	3
VI	(CWmin+1)/2 − 1	CWmin	2
VO	(CWmin+1)/4 − 1	CWmin	2

If the last frame was not received at this station with a correct MAC FCS value, the station defers transmission for EIFS − DIFS + AIFSN[AC] × slot time. The reason for using EIFS in the standard is the following. The virtual carrier sense mechanism indicates that the NAV at all stations is updated from the headers of all frames, not necessarily frames addressed to this station, nor frames from the same BSS. However, the header can only be interpreted from error-free frames, as ascertained by a correct value of the frame-check sequence (FCS). When activity on the wireless medium was detected, but a frame was not received correctly, the station knows that communication is under way. The station does not know the length of the period following the end of the frame. The EIFS is long enough to protect the acknowledgment from collision with a station that was not able to update its NAV. The backoff calculation uses the DCF method, with different contention window values for each queue. Similarly, the minimum and maximum contention window limits are not fixed per PHY, as with DCF, but are variable values, assigned to each traffic category either by a management entity or by the AP. In this way, each output queue contends for transmission opportunities.

At each transmission opportunity, a traffic scheduling entity selects a frame for transmission. If the backoff timers for more than one queue reach zero at the same slot, then the frame from the highest-priority queue is transmitted. These lower-priority queues will treat this as a collision and will set their contention window values as if they had experienced a transmit failure. The subtle difference is that retry bits are not set, as it would be done after an actual external collision on the wireless medium. The effect of the backoff procedure is that when multiple stations and/or queues of equal priority at one or more enhanced stations are deferring and go into random backoff, then the entity selecting the smallest backoff time using the random function will win the contention. In the case of queues of unequal priority at enhanced stations, queues with shorter AIFS[i] periods will win the contention. The higher-priority queue will lose the contention if its AIFS is longer. As stated previously, the Contention Window values, plus Short and Long Retry Counts are different for each queue i: CW[i], QSRC[i], and QLRC[i]. In the case of a collision the contention window increases in the same way as for the DCF. Provided that the contention window is less than the maximum, it is replaced by (CW[AC] + 1) × 2 − 1. Then out of this new, enlarged CW, another uniformly distributed backoff counter is drawn with a uniform distribution over the range [0,CW[AC]] inclusive, to reduce the probability of a new collision. The CW never exceeds the parameter CWmax[TC], which is the maximum possible value for CW. The CW is reset to CWmin after every successful attempt to transmit or when the long and short retry counters reach a certain limit. Assigning shorter contention windows to data with higher priority ensures that, on average, higher-priority data will get through before lower-priority data. A continuation of a transmission opportunity obtained the EDCA mechanism can be granted SIFS after a successful transmission, provided the device has another frame to transmit with the same AC and the duration of this frame plus any acknowledgment, if required, will fit in the remaining TXOP duration.

Note that to ensure fairness and overall improved QoS to all AC, 802.11e provides a mechanism for dynamic priority assignment. Each station is required to maintain two variables for every access category: used time and admitted time. At the time of association the admitted time is set to zero. Later the station can request admitted time for a specific priority level. If the used time reaches or exceeds the admitted time the station will no longer transmit data at the given priority level. However, the same data can still be transmitted with a priority levels Background or Best effort.

A traffic specification (TSPEC) describes the traffic characteristics and the QoS requirements of a traffic stream (TS). The main purpose of the TSPEC is to reserve resources within the HC and modify the HC's scheduling behavior. It also allows other parameters to be specified that are associated with the traffic stream, such as acknowledgment policy. TSPEC parameters not specified by the application are determined based on assumptions about throughput, latency, and packet error rate requirements.

The second mechanism for access to the wireless medium, which is part of the HCF, is called HCF-controlled channel access (HCCA). It provides the capability for reservation of TXOPs with the HC. A station can make requests with the HC for TXOPs for its own transmissions, as well as transmissions from the HC to itself. The HC either accepts or rejects the request based on admission control policy. The admission control policy depends on factors such as the vendors' implementation of the scheduler, available channel capacity, etc. This mechanism is expected to be used for the transfer of periodic traffic such as voice and video. If the traffic stream is admitted, then the scheduler will service the station during a service period (SP). This SP is a period of time during which one or more downlink frames are transmitted and one or more polled TXOPs are granted to the station. A station that is PS mode must wake-up before the start of the SP and may return to PS mode after the SP mode ends.

The scheduler is responsible for determining the mean data rate, packet size, and maximum service interval or delay bound. The schedule for an admitted stream is calculated in two steps. The first step is the calculation of the scheduled service period. In the second step, the TXOP duration for a given service period is calculated for the stream. In the first step, the scheduler finds the minimum of the maximum service periods for all admitted streams and then chooses a number lower than this minimum that is a submultiple of the beacon interval. This value is the service period for all stations with admitted streams. After the service period is calculated, the number of packets N_1 can be found from

$$N_1 = \frac{\text{mean data rate} \times SP}{L_1} \qquad \text{Eq. 2-2}$$

where L_1 is the chosen packet size. If the physical layer transmits at a data rate R, the transmission opportunity is the maximum to transmit N_1 packets of the chosen packet size or one packet of maximum size:

$$TXOP_i = \max\left(\frac{N_i \times L_i}{R} + O, \frac{L_{max}}{R} + O\right) \qquad \text{Eq. 2-3}$$

where L_{max} is the maximum packet size (2304 bytes). An example is shown in Figure 2–17. In this example the beacon interval is 100 ms and the maximum service period for the stream is 60 ms. As shown in the figure, the scheduler calculates a service period equal to 50 ms using the steps explained previously.

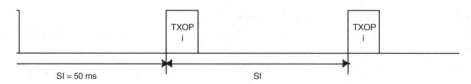

Figure 2–17:   Schedule for traffic stream *i*

The same process is repeated continuously while the maximum service period for the admitted stream is larger than current service period. Stream $k + 1$ is admitted if the calculated $TXOP_{K+1}$ satisfies the inequality

$$\frac{TXOP_{K+1}}{SP} + \sum_{i=1}^{K} \frac{TXOP_{K+1}}{SP} \leq \frac{T - T_{CP}}{T} \qquad \text{Eq. 2-4}$$

where L_{max} is the beacon interval and T_{CP} is the contention period. Otherwise, the $k + 1$-th stream is rejected. In this way, all admitted streams have access to the channel.

The HCCA can use polling by issuing QoS (+) CF-Polls to stations during the contention-free period. However, because the hybrid coordinator can also grant polled transmission opportunities by sending QoS (+) CF-Poll frames during the contention period, it is not mandatory for the hybrid coordinator to use the contention-free period for QoS data transfers. Stations receiving a QoS (+) CF-Poll are required to respond within a short IFS period. If the polled station has no queued traffic to send, or if the data available to send are all too long to transmit within the specified transmission opportunity limit, the station will send a QoS Null frame. In the case of no queued traffic, this QoS Null has a QoS control field that reports a traffic category queue size of 0. In the case of insufficient transmission opportunity size, this QoS Null has a QoS control field that reports the requested transmission opportunity duration needed to send the selected data and traffic category identifier for the highest priority data that is ready for transmission.

It is clear that there are two ways a station can obtain a transmission opportunity. If the station receives a QoS (+)CF-Poll during the contention period or contention-free period, or if it wins an instance of EDCF contention (which will only occur during the contention period). In the case of a polled transmission opportunity, the entire transmission opportunity is protected by the NAV set by the duration of the frame that contained the QoS (+)CF-Poll function.

Under the hybrid-coordination function, the basic unit of allocation of the right to transmit onto the wireless channel is the transmission opportunity. Each transmission opportunity is defined by a particular starting time, relative to the end of a preceding frame, and a defined maximum length. The novelty brought by IEEE 802.11e is that TXOPs not only have a starting time, but they also have a maximum duration. The duration is between 32 and 8160 μs. TXOPs are allocated by contention (EDCA-TXOP) or granted through HCCA (polled-TXOP). Every frame sent as a result of a polled transmission opportunity contains the transmission opportunity limit in its QoS control field. When polling is used in the original IEEE 802.11, only a single data frame can be transmitted in response. IEEE 802.11e allows, in response to a poll, the station to transmit as many data frames as will fit within the specified duration. These frame exchanges must be separated by a short IFS interval. Any station will not initiate transmission of a frame unless the transmission, and any acknowledgment or other immediate response expected from the peer MAC entity, can be completed before the end of the remaining duration of the transmission opportunity. However, data can be fragmented to fit in a transmission opportunity. Furthermore, no transmission opportunity nor transmission within it is allowed to extend across TBTT. Because this rule applies to transmission opportunities as well as transmissions within these transmission opportunities, it is the responsibility of the hybrid coordinator to ensure that the full duration of any granted transmission opportunity does not cross any of these boundaries. (See Figure 2–18.)

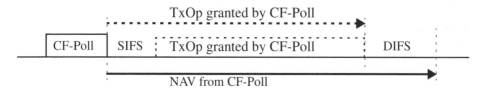

Figure 2–18: TXOP

There are certain rules about what type of data can be transmitted during a transmission opportunity. Management data and non-QoS data must be the only or the last frame exchange sequence in the transmission opportunity. Multiple control frames and QoS data frames can be sent in a single TXOP, with the nonfinal bit in all but the final frame set to 1 to indicate that more frames are coming. If an expected acknowledgment to a frame with nonfinal bit set to 1 is not received, the station awaiting the acknowledgment should transmit either its next frame or a QoS Null. All of these frames (management, control, QoS data, and non-QoS data) contain a duration field. The initial frame of a multiple frame exchange sequence contains a duration value, which is the remaining duration of the TXOP. The initial frame of the sole or final frame exchange sequence contains a duration value that covers the actual remaining time needed to transmit the frame, plus one ACK time, plus one SIFS interval, plus one DIFS interval. This is important so that all stations can maintain their NAVs.

During the CP, each TXOP begins either when the medium is determined to be available under the EDCA rules, i.e., after AIFS plus backoff time, or when the station receives a special poll frame, the QoS CF-Poll, from the HC. The QoS CF-Poll from the HC can be sent after a PIFS idle period without any backoff. However, although the poll frame is a new frame as part of the upcoming 802.11e, the legacy stations also set their NAVs upon receiving this frame. Therefore, the HC can issue polled TXOPs in the CP using its prioritized medium access. During the CFP, the starting time and maximum duration of each TXOP is specified by the HC, again using the QoS CF-Poll frames. Stations will not attempt to get medium access on their own during the CFP, so only the HC can grant TXOPs by sending QoS CF-Poll frames. The CFP ends after the time announced in the beacon frame or by a CF-End frame from the HC.

The HC may allocate TXOPs to itself to initiate fame deliveries whenever it wants, but only after detecting the channel as being idle for PIFS. The duration values used in the QoS frame exchange reserve the medium for DIFS longer than the end of the sequence to permit continuation of a NAV-protected contention-free transfer by the concatenation of several contention-free blocks. This extra reservation of the wireless channel allows the hybrid coordinator to initiate a subsequent transmission opportunity with reduced risk of collision because stations cannot begin contending for the medium until DIFS after the last communication within the transmission opportunity.

The hybrid coordinator performs delivery of queued broadcast and multicast frames following DTIM (Delivery Traffic Indication Message) beacons in a contention-free period. The hybrid coordinator may use a longer contention-free period for QoS delivery and/or QoS polling by continuing with hybrid-coordination function frame exchange sequences, after broadcast/multicast delivery, for a duration not exceeding the maximum possible duration of the contention-free period. The HC may also operate as a PC, providing (non-QoS) CF-Polls to stations.

Because the hybrid coordinator is a type of point coordinator, the hybrid coordinator includes a contention-free Parameter Set element in the Beacon frames it generates. This causes the BSS to appear to be a point-coordinated BSS. This causes all stations (other than the hybrid coordinator) to set their NAVs to the corresponding duration value at TBTT. This prevents most contention with the contention-free period by preventing nonpolled transmissions by all stations whether or not they are CF-Pollable.

In this way, all stations inherently obey the medium access rules of the hybrid-coordination function.

When a station updates its NAV setting using the duration value from a QoS CF-Poll containing the BSSID of this BSS, that station will also save the address of the holder of the transmission opportunity. If an RTS, management type, data type or QoS data type frame is received from a MAC address that matches this saved TXOP holder address, the station will send the appropriate response after SIFS, without regard for—and without resetting—its NAV. This saved TXOP holder address is cleared whenever the NAV is reset, the NAV counts down to 0, or when a QoS data type frame is received with an address that matches the saved TXOP holder address and has the non-final bit set to 0. The NAV is cleared when the station receives a CF-End frame or a QoS CF-Poll with a duration value equal to zero.

During the CP, the NAV can be reset in the following manner. When a QoS data frame is received from the TXOP holder, and the non-final bit in the QoS control field is set to 0, and the No ACK bit is set to 1, the NAV is reset. During the CP, when everything is the same, except that the No ACK bit in the QoS control field is set to 0, the NAV is reset.

During the CP, if an enhanced station receives a QoS data type frame that requires acknowledgment, when its NAV is set and the saved TXOP holder address is clear, the response after SIFS is an ACK frame, even if the frame being acknowledged was of a subtype including QoS CF-Poll. In this manner, the responding enhanced station indicates that it is unable to accept the TXOP conveyed by the QoS CF-Poll.

Another way to achieve improved NAV protection during either the CP or the CFP is to use the well-known RTS/CTS exchange. The sending of RTS during the CFP is used not to update the NAV, which is protected for the entire duration of the CFP. RTS still can be used to ensure that the addressed recipient station is within range and awake, and to protect against hidden nodes. Stations can send an RTS as the first frame of any frame exchange sequence without regard for RTS frame threshold. The corresponding CTS will set the NAV in the vicinity of the receiver, and its duration field is set to the number of microseconds until the end of the TXOP.

The HC is a type of point coordinator, but differs from the point coordinator used in PCF in several significant ways. Most important is that the hybrid-coordination function frame exchange sequences may be used among stations during both the contention period and the contention-free period. Another significant difference is that QoS CF-Poll grants transmission opportunities with duration specified in the QoS (+) CF-Poll frame. As a result, stations can set their NAV appropriately, which is not possible if the duration is not specified. Stations also can transmit multiple frame exchange sequences within a given transmission opportunity. The hybrid coordinator not only delivers queued broadcast and multicast data, but can use a longer contention-free period for QoS delivery and/or QoS polling by continuing with hybrid-coordination function frame exchange sequences after broadcast/multicast delivery. Finally, the HC may also operate purely as a PC, providing (non-QoS) CF-Polls using all the applicable rules for the PCF.

It is interesting to consider the coexistence of the DCF, PCF, and HCF. The DCF and a point coordination function (either PCF or HCF) coexist in a manner that permits both to operate concurrently within the same network. When a point coordinator is operating in a basic service set, the point coordination function and the distributed coordination function access methods alternate, with a contention-free period followed by a contention period. When an HC is operating, there is a contention-free period and a contention period in each superframe, and stations treat the hybrid coordinator as if it were a point coordinator, using the distributed coordination function access method only during the contention period. The hybrid-coordination function access methods (polled and contention-based) operate concurrently, throughout the superframe. Concurrent operation allows the polled and contention-based access methods to alternate within intervals as short as the time to transmit a pair of frame exchange sequences. Note that it is possible to use both PCF and HCF in a single contention-free period. This, however, is extremely complex and does not seem advantageous.

The hybrid coordinator is always implemented in the access point. In an IBSS, there is no access point, no hybrid coordinator, and therefore no polled transmission opportunities. In an IBSS, QoS is supported only by the EDCA. In an IBSS priorities and differentiation of traffic classes are static parameters.

Since the QoS facility was standardized recently, there will be a lot of stations that do not support it. An interesting situation is when, in a basic service set, some of the stations support the QoS facility and some do not. Because the hybrid coordinator is a type of point coordinator, the hybrid coordinator includes a contention-free Parameter Set element in the Beacon frames it generates. This causes a BSS with enhanced stations to appear to be a point-coordinated BSS to legacy stations. This causes all stations (other than the hybrid coordinator) to set their NAVs to the corresponding duration value. This prevents most contention with the contention-free period. The HCF assumes that all non-QoS data has priority equivalent to best effort.

It should also be noted that real-time data does not require low-bit error rate like computer data. In particular if a voice or video packet has a few bit errors, then its quality is usually not affected significantly. However, at present this characteristic is not used by IEEE 802.11. The MAC will not pass to the higher layers any packets that contain bit errors. Research work in this area is likely to continue.

Some applications require timers at different stations to be synchronized. To support these applications the standard defines a MAC service. One way to accomplish synchronization across a BSS is by multicasting synchronization. In this method synchronization packets from the higher layers containing a time stamp and a sequence number are broadcasted. These packets can be recognized by their MAC headers. The time stamp in the synchronization packet would contain the higher layer clock value at the time when the previous synchronization packet was transmitted. The sequence number would include a value corresponding to the packet for which the time stamp is provided.

POWER-SAVING MECHANISM

The 802.11 MAC allows stations to enter a low-power mode. While in power-save (PS) mode, a device alternates between awake and doze states. The doze state is a power-efficient state, and the device is in the awake state only for very short periods of time.

In a BSS, the station will first inform the AP when it is going to enter power-saving mode and the number of beacon periods that it will be in this mode. The AP is responsible then to buffer information for this station while it is in doze state. When the station returns to the awake state, these frames are delivered using the EDCA or HCCA mechanisms. The PS device will wake up just to receive a beacon, but it does not have to wake up for every beacon. The beacon contains information whether the AP has packets queued for it. If there are queued packets, the PS device will remain in the awake state and contend for the medium to send a poll frame to the AP. Upon receipt of a poll frame, the AP will contend for the medium to send these buffered packets. The contention-free period can also be used to deliver buffered frames. Service periods in response to a poll are called unscheduled service periods. The device must also awaken when multicast frames are to be delivered. These times are indicated in the beacon as the traffic indication map (TIM). The delivery TIM (DTIM) count is an integer value that is equal to the number of beacon frames that will occur between the delivery times of multicast frames. In this way power-saving stations do not have to wake-up for every beacon. A larger value of the DTIM count leads to greater power efficiency, but also to increased latency for multicast traffic.

To improve the QoS that is provided by the PS mechanism 802.11e introduced the notion of scheduled and unscheduled service periods. A scheduled service period is a service period that begins at a scheduled wake-up time. A scheduled SP starts at fixed intervals of time determined by the minimum service interval.

A station may be in power-save mode before the set up of DLP or block acknowledgment. Once DLP is set up with another station, the power save mode is suspended. In a BSS, block acknowledgment mode is suspended during the PS mode. In an IBSS, stations are required to suspend their power-save mode if they have set up block acknowledgment connections.

In an IBSS, a station that wants to enter power-saving mode must inform at least one other station that it is about to do so by setting appropriate power management bits within the header of transmitted frames. All stations in an IBSS before transmitting must estimate the power-saving state of the destination. If the destination is deemed to be in power-saving state, they should refrain from transmitting and buffer the traffic. Then the power-saving device must wake up to receive every beacon. The device must also stay awake for a certain period of time after every beacon, called ad-hoc traffic indication message (ATIM) window. During this time period, other stations will make an announcement if they have data for the power-saving station. Multicast information will also be announced during this time period. If the power-saving station receives an ATIM frame, it must acknowledge the ATIM frame and remain awake to allow the other stations to deliver their data destined for it. Clearly, the power savings in an IBSS are less than the savings in a BSS, because the station is required to wake up at every beacon and remain awake for a certain period of time after the beacon.

IEEE 802.11 PHYSICAL LAYERS

The protocol data unit for all physical layers consists of a preamble and a header, followed by MAC data. The receiver uses the preamble for detection and synchronization. The header has information about the MAC data such as data rate and length. If the physical layer supports multiple data rates the preamble and the header are sent at the lowest data rate. In this way, the preamble and the header are sent in a way to ensure maximum range. The MAC frame that follows consists of the following basic components:

a) A header, which comprises frame control, duration, address, sequence control information, and, for QoS frames, QoS control information

b) A variable length frame body

c) A frame check sequence (FCS), which contains an IEEE 32-bit cyclic redundancy code (CRC).

This CRC protects all fields in the MAC header and frame body. Using the polynomial

$$G(x) = x^{32} + x^{26} + x^{23} + x^{22} + x^{16} + x^{12} + x^{11} + x^{10} + x^8 + x^7 + x^5 + x^4 + x^2 + x + 1$$

the 32-bit CRC is the ones complement of the sum modulo 2 of the following two quantities. The first quantity is the remainder of the division modulo 2 of $x^k(x^{31} + x^{30} + \ldots + x + 1)$ by $G(x)$, where k is the total number of bits in the protected fields, the MAC header and frame body. The second quantity is the remainder of the division of the content of the protected fields treated as a polynomial and multiplied by x^{32}, by $G(x)$. The CRC is an error-detection mechanism. At the receiver, the MAC passes to the higher layers only frames for which the received and calculated CRC values match.

IEEE 802.11 has at present six physical layers: three physical layers were defined initially, and three were developed later. The first three physical layers are direct-sequence spread-spectrum, frequency-hopping spread-spectrum, and infrared. They provide data rates of 1 and 2 Mb/s. The frequency-hopping physical layer uses two-level and four-level Gaussian frequency shift keying for transmission at 1 Mb/s and 2 Mb/s, respectively. In the 2.4 GHz band, the total number of 1-MHz-wide hop channels is 79, from 2402 MHz to 2480 MHz. These 79 channels are divided into three nonoverlapping sets. Frequency-hopping radios must hop at least at a rate of 2.5 hops per second with a minimum hop distance of 6 MHz. The minimum hop separation ensures that if one hop falls on a "bad" frequency (with a low signal-to-noise ratio), then the next hop will likely be on a "good" frequency. The frequency-hopping physical layer was used in some early products, but because the data rate is only 1 Mb/s or 2 Mb/s, the overall market was relatively small. The infrared physical layer is based on pulse-position modulation (PPM), achieves data rates of 1 Mb/s and 2 Mb/s, and requires line-of-sight. The frequency-hopping and infrared physical layers are not relevant at present because of their performance. The DSSS physical layer is considered next.

DSSS PHY

The DSSS physical layer operates in the 2.4 GHz band. The protocol data unit for this physical layer, like for all physical layers, consists of preamble, header, and MAC data. The preamble contains a 128-bit SYNC word, followed by a 16-bit start-of-frame delimiter. The preamble length in bits is 144 bits, and in time is 144 μs. Following the preamble, the header identifies what type of modulation is used for the MAC frame, the length of time required to transmit the MAC frame (measured in microseconds), and, as an error-detection mechanism, a 16-bit CRC. The 16-bit CRC is the ones complement of the remainder of the modulo-2 division of the contents of the header by the polynomial $x^{16} + x^{12} + x^5 + 1$. The 16-bit CRC is calculated before scrambling. If the 16 bits of the CRC match at the receiver, the PHY header has been correctly received. The preamble and the header are always sent at 1 Mb/s. The MAC data that follows can be transmitted at

1 Mb/s or 2 Mb/s. The modulation that the original standard supports is differential phase-shift keying.

Two types of differential phase shift keying are supported. The preamble and the header are transmitted using DBPSK. The MAC data is sent using either DBPSK, resulting in a data rate of 1 Mb/s, or using DQPSK resulting in a data rate of 2 Mb/s. The bit-to-symbol mapping for DBPSK is that 0 is encoded with no phase change, and 1 is encoded with a phase change of 180 degrees. For DQPSK the input bits are grouped in pairs of two. Then 00 is encoded with no phase change, 01 with a phase change of 90 degress rotating counterclockwise, 11 is encoded with a phase change of 180 degrees, and 10 of 270 degrees. Noncoherent demodulation can be used because a clock reference is not needed to recover the data. DBPSK is more robust compared with DQPSK. DBPSK achieves the same bit error rate at a lower signal-to-noise ratio compared with DQPSK.

The signal processing steps performed are scrambling, spreading, and modulation. Scrambling randomizes the data using the polynomial $1 + z^{-4} + z^{-7}$. The spreading method employs the 11-bit Barker sequence 1,−1,1,1,−1,1,1,1,−1,−1,−1. This Barker sequence is applied to a modulo-2 adder together with each of the information bits. The chipping rate is 11. The output of the adder is a signal, the data rate of which is 11 times higher than the information rate. This ensures a spreading factor and equivalently processing gain of 10.8 dB. The signal bandwidth becomes 22 MHz. Because the total bandwidth available is 83.5 MHz, with appropriate guard bands, there are three nonoverlapping 22-MHz-wide channels in the band. The first channel is centered at 2.412 GHz, the second is centered at 2.437 GHz, and the third is centered at 2.462 GHz. The maximum allowed output power in North America is 1W; however, many vendors have selected 100 mW as the default transmit power level. The reasons are that lower transmit output power does not drain batteries as fast, and 100 mW is the maximum output power per European regulations. Also, this decreases the range of operation of the access points, enabling vendors to sell more higher-margin access points.

IEEE 802.11b

IEEE 802.11b is discussed before IEEE 802.11a, because IEEE 802.11b is still based on spread-spectrum technology. IEEE 802.11b achieves data rates of

5.5 Mb/s and 11 Mb/s. Most of the products on the marketplace today are IEEE 802.11b. How does the IEEE 802.11b PHY work? Similarly to all physical layers a frame consists of a preamble, header, and MAC data. The preamble is used by the receiver to detect the signal and to synchronize with the transmitter. There are two preambles: long and short. The long preamble is identical to the IEEE 802.11 preamble described in the previous section. Because the long preamble provides backwards compatibility with IEEE 802.11, it is mandatory to implement. The short preamble is optional.

Two preambles are defined because the long is transmitted at 1 Mb/s, and therefore transmission of the preamble will take a significant amount of time. Equivalently, devices will be in a low-data-rate mode for a long time and will not take full advantage of the higher data rates that IEEE 802.11b provides. The short preamble helps to increase the actual throughput. This is an example of the trade-off between backwards compatibility and throughput, which is even more important for the other extensions of IEEE 802.11. If an IEEE 802.11b device uses the long preamble (which would be understood by legacy IEEE 802.11 stations) then the throughput is lower. If the long preamble is used, then the header is also according to IEEE 802.11. Legacy stations will be able to decode the header, and in particular will decode the information about the length of the MAC data that follows. From the length information, they can update their NAVs and can maintain silence during the transmission. If the long preamble is used, the MAC data that follows can be transmitted at 1, 2, 5.5, and 11 Mb/s. If the short preamble is used, the throughput would be higher, but legacy IEEE 802.11 equipment cannot decode the preamble. The short preamble uses a 56-bit SYNC field and is transmitted also at 1 Mb/s using DBPSK modulation and spreading using the 11-bit Barker code. If the short preamble is used, however, the header is transmitted at 2 Mb/s using DQPSK and Barker spreading. The MAC data that follows the header is transmitted at 2, 5.5, or 11 Mb/s. The actual data rate used is denoted in the header.

All transmitted bits are scrambled and descrambled using the polynomial $1 + z^{-4} + z^{-7}$, which is the same as for 802.11. The four modulation formats and data rates defined in IEEE 802.11b are: (1) the basic rate of 1 Mb/s modulated with DBPSK; (2) the extended rate of 2 Mb/s with DQPSK modulation; (3) and (4) two enhanced rates of 5.5 Mb/s and 11 Mb/s using CCK modulation. As an option alternative to CCK, PBCC can be used. The Barker code is the spreading code for the basic and extended rates. The enhanced rates of 5.5 and 11 Mb/s do not use the Barker code. CCK modulation was chosen by the IEEE 802.11 Working Group, because it helps achieve backwards compatibility by maintaining the same 22 MHz-wide channels. CCK is a variation of M-ary orthogonal keying modulation. Spreading is achieved by using a spreading code with eight samples. Each of these eight chips is complex and obtained using QPSK modulation. The eight chips of the spreading code are

$$\{e^{j(\varphi_1+\varphi_2+\varphi_3+\varphi_4)}, e^{j(\varphi_1+\varphi_3+\varphi_4)}, e^{j(\varphi_1+\varphi_2+\varphi_4)},$$
$$-e^{j(\varphi_1+\varphi_4)}, e^{j(\varphi_1+\varphi_3+\varphi_4)}, e^{j(\varphi_1+\varphi_3)}, -e^{j(\varphi_1+\varphi_2)}, e^{j\varphi_1}\}$$

These eight values are obtained from a Walsh/Hadamard code, where φ_1 is added to all samples, φ_2 is added to every other sample, starting with the first, φ_3 is added to every other group of two of samples, and φ_4 is added to every other group of four samples. The fourth and seventh sample are multiplied by -1, or equivalently rotated by 180 degrees to optimize the correlation properties and minimize the DC offset. Walsh/Hadamard codes are also called complementary code and this is why the name CCK is used. The spreading code has eight samples, but the effective chipping rate is must be 11 to maintain the same channel bandwidth of 22 MHz. To achieve this the symbol rate is increased to 1.375 Msymbols/s.

For transmission at 5.5 Mb/s, four bits $d_0d_1d_2d_3$ are transmitted per symbol. The first two bits d_0d_1 encode φ_1 based on DQPSK—φ_1 is 0, 90, 180, or 270 degrees relative to the preceding symbol, if d_0d_1 are 00, 01, 11, or 10, correspondingly. In addition, starting with the second symbol, every other symbol is multiplied by -1. This multiplication by -1 can be represented by 180 degrees being added to φ_1 for the second, fourth, sixth, etc., symbols after the header. The other two bits d_2 and d_3 determine $\varphi_2 = d\pi_2 + \frac{\pi}{2}$ and $\varphi_4 = d_3\pi$, while $\varphi_3 = 0$.

For transmission at 11 Mb/s, eight bits $d_0d_1d_2d_3d_4d_5d_6d_7$ are transmitted per symbol. The first two bits d_0d_1 encode φ_1 exactly in the same way as for transmission at 5.5 Mb/s, with 180 degrees added to φ_1 for the second, fourth, sixth, etc., symbols after the header. The pairs d_3d_4, d_4d_5, and d_6d_7 encode φ_2,

φ_3, and φ_4, based on QPSK. These phases are 0, 90, 180, or 270 degrees counterclockwise, if the corresponding bit pair is 00, 01, 10, or 11, respectively.

IEEE 802.11b also defines an optional coding scheme called PBCC. It sometimes is referred to as high-performance mode, although it achieves the same data rates. It uses a 64-state rate one-half binary convolutional code and a cover sequence. BPSK is used for 5.5 Mb/s and QPSK for 11 Mb/s.

A PBCC modulator uses a 64-state rate-1/2 binary convolutional code (BCC) and a cover sequence, as shown in Figure 2–19.

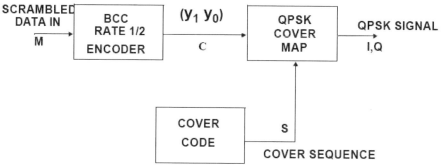

Figure 2–19: PBCC modulator

The binary convolutional encoder is shown in Figure 2–20. After the convolutional encoding, bit-to-symbol mapping is performed. For transmission at 5.5 Mb/s, the bit-to-symbol mapping is performed using BPSK. For transmission at 11 Mb/s, it is performed using QPSK.

The mapping from the output bits of the binary convolutional encoder to the BPSK and QPSK symbols is determined using a pseudorandom cover sequence. QPSK modulation using the pair of bits $y_1 y_0$ is done as shown in Figure 2–21 and BPSK modulation as shown in Figure 2–22.

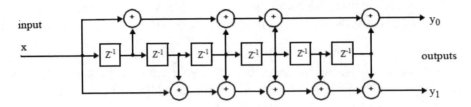

Figure 2–20: Binary convolutional encoder

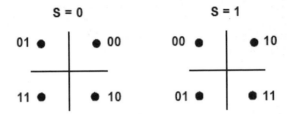

Figure 2–21: Cover code bit-to-symbol mapping in QPSK mode

Figure 2–22: Cover code bit-to-symbol mapping in BPSK mode

The cover sequence has 256 bits, generated using the 16-bit seed sequence 0011001110001011. The first 16 bits of the cover sequence are identical to the seed sequence. To obtain the second, third, etc. groups of 16-bits in the cover sequence, the seed sequence is cyclically shifted on the left by 3, 6, etc. Every bit in the cover sequence determines how the bit-to-symbol mapping is performed for every bit in the frame. For frames that have more than 256 bits, the cover sequence is simply repeated.

IEEE 802.11b uses the same frequency channels as 802.11. The channels center frequencies are spaced 25 MHz apart to allow multiple WLANs to operate simultaneously in the same area without interfering. The North American channelization scheme is shown in Figure 2–23 and the European scheme is shown in Figure 2–24.

The range of operation of 802.11b is 150 feet for a floor divided into individual offices by concrete or sheet-rock, about 300 feet in semi-open indoor spaces such as offices partitioned into individual workspaces, and about 1000 feet in large open indoor areas.

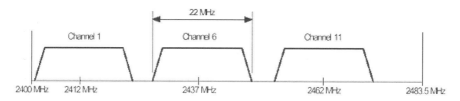

Figure 2–23:   North American channelization for 802.11b

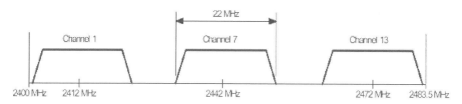

Figure 2–24:   Nonoverlapping 802.11b channelization for Europe

IEEE 802.11a

As a result of the Telecommunications Act of 1996, in 1997 the FCC allocated bands in the 5 GHz region for unlicensed use [B41], [B42]. These bands are in total 300 MHz wide. As discussed in "Government regulations" on page 5, in 2003 the FCC made also available the band between 5.47 and 5.725 GHz.

This chunk of spectrum allows significantly higher speeds of communication than the 83.5-MHz-wide ISM band between 2.4 GHz and 2.4835 GHz. At the same time, a technology called orthogonal frequency division multiplexing (OFDM) became very well developed, widely understood, and economically feasible for incorporation in mass-market wireless products—all the necessary characteristics for inclusion in a standard. In 1997, IEEE 802.11 seized the opportunity to create a powerful high-speed physical layer in the newly allocated UNII bands. This standard, IEEE 802.11a, was developed in a very short time—less than two years [B69]. At the present IEEE 802.11a products are available on the market.

Since the IEEE 802.11a PHY is using OFDM and not spread spectrum technology, it is worth comparing OFDM and spread spectrum. In high-data rate wireless communications, OFDM is superior for several reasons. The main reason is that high-data rates cannot be achieved with spread spectrum technology because the spectrum is limited. The benefits of spread spectrum are proportional to the spreading factor. However, in a limited band the spreading factor cannot be very high for a high data rate. OFDM is the technology of choice for high-data rate wireless communications. In OFDM the transmission bandwidth is divided into many narrow subchannels, which are transmitted in parallel. Ideally each subchannel is narrow enough so that the fading it experiences is flat. OFDM allows the subchannels to overlap, which is spectrally efficient. Since the subchannels overlap, how are they separated at the receiver? The subchannels can be separated at the receiver because they are orthogonal. OFDM combats ISI caused by multipath by using symbol periods several times longer than the delay spread of the channel impulse response.

Like the other physical layers, the OFDM PHY takes the MAC data and adds a preamble and a header. The preamble is used for detection, synchronization, automatic gain control (AGC) estimation, coarse and fine carrier frequency estimation, and channel estimation. The preamble consists of 10 short training symbols, each 0.8 μs long, and two long training symbols, each 3.2 μs. The long training symbols are preceded by a long guard interval of 1.6 μs. This guard interval is a cyclic prefix of a long training symbol. The total preamble is 16 μs long and is sent using BPSK and OFDM using convolutional coding rate of one-half. A header follows the preamble. The header contains information about the MAC data that follows. The header first has information about the data rate, the length of the transmission, and parity and tail bits. The signal field is transmitted using BPSK and OFDM using convolutional coding rate of one-half.

The following signal processing steps are performed: First, the data is scrambled for randomization. The scrambler is defined by the polynomial $x^7 + x^4 + 1$, (Figure 2–25) and is identical with the descrambler.

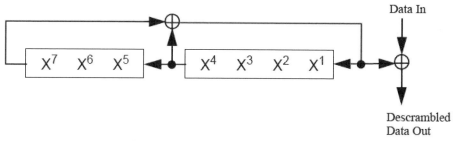

Figure 2–25: Scrambler circuit

The scrambler is initialized at the transmitter with a pseudorandom number. To provide the receiver with the capability to estimate the initial state of the scrambler, the seven least-significant bits of the header are initialized with zeros and then scrambled.

The bits after scrambling are encoded by a convolutional code of rate 1/2, 2/3, or 3/4. The rate-1/2 convolutional encoder is described by the polynomials 133 and 171 in octal notation (Figure 2–26). The higher data rates are obtained by puncturing omission of certain bits in the codeword. Decoding is performed by a hard-decision, or even better, soft-decision Viterbi decoder.

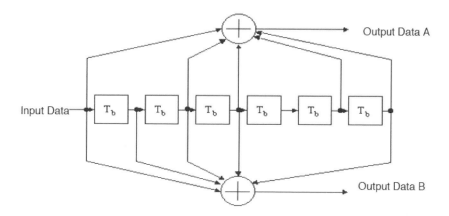

Figure 2–26: Rate one-half convolutional encoder

After convolutional encoding, the encoded bits are interleaved using a block interleaver with a block size corresponding to the number of bits in one OFDM symbol. The interleaver is defined by a two-step permutation. The first permutation ensures that adjacent coded bits are mapped onto nonadjacent subcarriers. The second ensures that adjacent coded bits are mapped alternately onto less and more significant bits of the constellation.

After interleaving, OFDM modulation is performed in the following way. From the interleaved bit sequence, complex symbols are obtained using BPSK, QPSK, 16 QAM, or 64 QAM modulation. The Gray-coded bit-to-symbol mappings are shown in Figure 2–27, Figure 2–28, and Figure 2–29. The value $(I + jQ)$ in these figures is multiplied by a normalization factor to ensure the same average power for all modulations. The normalization factor is 1 for BPSK, $1/\sqrt{2}$ for QPSK, $1/\sqrt{10}$ for 16-QAM, and $1/\sqrt{42}$ for 64-QAM. These complex symbols in groups of 48 become the subcarriers of one OFDM symbol. Four subcarriers in fixed locations are reserved as pilot tones, bringing the total number of nonzero subcarriers 52. In addition, 12 subcarriers are equal to zero, which effectively introduces a guard frequency band. The total number of subcarriers is 64, which allows a 64-point inverse fast Fourier transform (IFFT) to be used. In practice, the IFFT hardware is implemented using a radix-2 algorithm, or the computationally more efficient radix-4 algorithm.

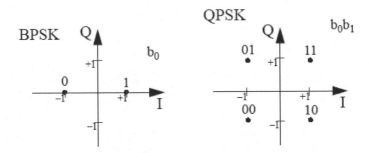

Figure 2–27:   Bit-to-symbol mappings for BPSK and QPSK

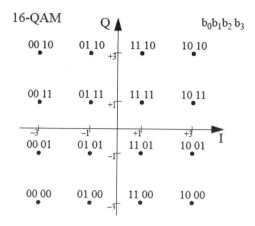

Figure 2–28: Bit-to-symbol mapping for 16-QAM

After the IFFT, a circular extension of the waveform is added as a cyclic prefix. Why is a cyclic prefix necessary? It is necessary to make linear convolution identical to circular convolution. The transmitted signal and the channel impulse response are linearly convolved. However, the product of the discrete Fourier transforms (DFT) of the transmitted signal and the channel impulse response corresponds to circular convolution. From digital signal processing, it is well known that to make circular convolution identical to linear convolution, a signal extension of appropriate length is necessary. The length of the cyclic prefix should be at least equal to the length of the channel impulse response. Without cyclic prefix the inter-symbol interference between adjacent OFDM symbols (e.g., the inter-block interference) is significant and will reduce the performance. The cyclic prefix resolves the inter-block interference problem. For 802.11a the cyclic prefix is 0.8 μs and therefore the inter-symbol interference will be reduced when the energy of the channel impulse response is contained within 0.8 μs.

If $r(t)$ is the complex baseband signal, the actual transmitted signal is

$$r_{RF}(t) = \text{Re}\{r(t)e^{j2\pi f_c t}\}$$

Eq. 2–5

where f_c is the carrier center frequency. The complex baseband signal, obtained using an IFFT is

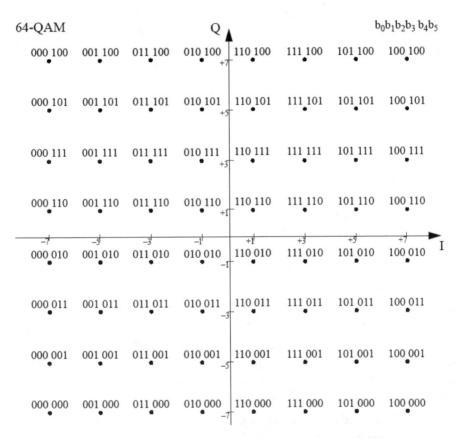

Figure 2-29: Bit-to-symbol mapping for 64-QAM

$$r(t) = w(t) \sum_{k=-26}^{k=26} C_k e^{j\frac{2\pi kt}{T_{FFT}}}$$ Eq. 2–6

where C_k are the complex symbols, obtained using BPSK or QAM modulation, and $T_{FFT} = 3.2$ μs is the FFT period, or the symbol period without the cyclic prefix. If the complex baseband signal is sampled with a sampling interval of T_{FFT} / N, or alternatively 3.2 μs/64 = 50*ns*, then it can be represented as

$$r(nT_{FFT}/N) = w(nT_{FFT}/N) \sum_{k=-26}^{k=26} C_k e^{j\frac{2\pi kn}{N}}$$ Eq. 2–7

or alternatively using discrete-time notation

$$r(n) = w(n) \sum_{k=-26}^{k=26} C_k e^{j\frac{2\pi kn}{N}}$$ Eq. 2–8

This makes OFDM modulation equivalent to an inverse DFT and allows FFT algorithms to be used. $w(t)$ is a window function. Two window functions are specified: rectangular window having values of 1 for every sample, and a window function, having value of 1 for every sample, except the first and the last sample, where the value is 0.5. The pilot tones are in carriers –21, –7, 7, and 21. Therefore, the contributions of the pilot subcarriers for the *n*-th OFDM symbol is obtained by an IFFT of $P_{-26,26}$, which is zero everywhere, except $P_{-26} = P_{-7} = P_7 = P_{21}$. To prevent the generation of spectral lines, the pilot carriers are BPSK modulated by pseudo-random binary sequence consisting of 1's and –1's. This pseudo-random binary sequence can be generated using the same scrambler in Figure 2–25, when the all-ones initial state is used, and all 1's are replaced with –1's and all 0's, are replaced with 1's. Every element from this pseudo-random sequence is used to obtain the four pilots of one OFDM symbol.

The short training symbol

$$r_{short}(t) = w(t) \sum_{k=-26}^{k=26} S_k e^{j\frac{2\pi kt}{T_{FFT}}} \qquad \text{Eq. 2-9}$$

is obtained by an IFFT of the sequence $S_{-26,26}$, which is zero everywhere except $S_{-24} = S_{-16} = S_{-4} = S_{12} = S_{16} = S_{20} = S_{24} = \sqrt{13/6}(1+j)$ and $S_{-20} = S_{-12} = S_{-8} = S_4 = S_8 = \sqrt{13/6}(-1-j)$.

One long training symbol

$$r_{long}(t) = w(t) \sum_{k=-26}^{k=26} L_k e^{j\frac{2\pi kt}{T_{FFT}}} \qquad \text{Eq. 2-10}$$

is obtained using an IFFT of $L_{-26,26}$ = {1, 1, −1, −1, 1, 1, −1, 1, −1, 1, 1, 1, 1, 1, 1, −1, −1, 1, 1, −1, 1, −1, 1, 1, 1, 1, 0, 1, −1, −1, 1, 1, −1, 1, −1, 1, −1, −1, −1, −1, −1, 1, 1, −1, −1, 1, −1, 1, −1, 1, 1, 1, 1}. To provide an opportunity for improved channel estimation accuracy, two long training symbols are used, preceded by a cyclic prefix. The preamble is the concatenation of the short and long training symbols.

A 64-point IFFT can be used. This requires that complex QAM symbols 1 to 26 are mapped to IFFT inputs 1 to 26, and complex QAM symbols −26 to −1 are mapped to IFFT inputs 38 to 63.

The data rates that 802.11a achieves are 6, 9, 12, 18, 24, 36, 48, and 54 Mb/s. One OFDM symbol is sent every 4 μs, of which 0.8 μs is the cyclic prefix. Since 250,000 symbols are sent every second, and one symbol uses 48 data carriers, a data rate of 6 Mb/s is achieved with BPSK modulation and a convolutional code of rate one-half (24 × 250,000 = 6 Mb/s). Data rate of 54 Mb/s is achieved with 64 QAM and a convolutional code rate of ¾.

Note that the slot time for 802.11a is required to be 9 μs. The slot time is the sum of the Rx-to-Tx turnaround time, MAC processing delay, and the time to perform clear channel assessment (CCA). The Rx-to-Tx turnaround time should be under 2 μs, the MAC processing delay should also be under 2 μs, and the CCA time should be under 4 μs.

The bandwidth of one OFDM symbol is 20 MHz and since there are 64 carriers one carrier takes about 20 MHz/64, or 312 KHz. Since only 52 carriers are used, the power density can be calculated assuming that one symbol occupies 16 MHz.

Therefore, output power of 200 mW, or 23 dBm, corresponds to power density of 12.5 mW/MHz, which is equal to 11 dBm/MHz and −49 dBm/Hz. The total bandwidth in the UNII bands is 300 MHz, but there are guard bands. In each of the three UNII bands four non-overlapping 20 MHz channels can be used. Therefore the total number of different OFDM channels is 12. Immediately one sees that besides high speed OFDM has another advantage. Since there are 12 different OFDM channels compared with only three different 802.11b channels, the total capacity of 802.11a is much higher.[9] (See Figure 2–30 and Figure 2–31.)

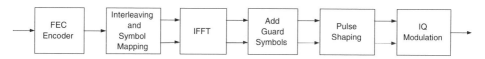

Figure 2–30: IEEE 802.11a transmitter according to the standard

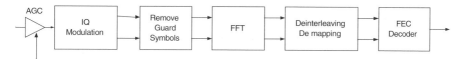

Figure 2–31: An implementation of a IEEE 802.11a receiver

Recently the IEEE 802.11 working group completed work on an amendment to IEEE 802.11a, called IEEE Std 802.11h. IEEE 802.11h defines how IEEE 802.11a devices implement dynamic frequency selection (DFS) and transmitter power control (TPC). IEEE 802.11h is necessary for two reasons. First, DFS and TPC are necessary is to satisfy European regulations, which require all unlicensed devices in the 5 GHz bands to implement DFS and TPC. Because the original IEEE 802.11a does not have DFS and TPC, IEEE 802.11a devices could not be sold in Europe. The second reason is that in the 5 GHz bands there are other wireless devices, such as satellite communication systems and radars, which are the primary users of the bands.

According to government regulations, these devices have the right-of-way, i.e., IEEE 802.11a must not cause interference to them. Anticipating widespread

---

[9] The capacity becomes even higher if the very recently allocated band of 5.47–5.725 GHz is also included.

adoption of WLAN, the radar community was concerned about interference from IEEE 802.11a. According to IEEE 802.11h, IEEE 802.11a devices must periodically test for the presence of radars. If radar is found in the 20-MHz-wide channel in which IEEE 802.11a is operating, the IEEE 802.11a devices must vacate the channel within a certain period of time. To accomplish this channel transfer, the IEEE 802.11a devices must find another channel and exchange messages to coordinate the channel transfer. If an AP is present, the AP decides which channel to transfer to. If an AP is not present, one of the stations assumes the role of DFS owner and it makes the decision as to which channel to transfer to. The channel switch announcement is made not according to the CSMA/CA rules of the DCF, but according to the PCF. In this way, IEEE 802.11h tries to satisfy the demands of the radar community by ensuring that the channel will be vacated as quickly as possible.

The second mechanism specified by IEEE 802.11h is TPC. TPC is a mechanism according to which the output power is adapted to regulatory constraints and channel conditions. Notice first that the maximum allowed output power levels in the lower, middle, and upper UNII bands are different. Furthermore, government regulations around the world are different. Therefore, devices can change their output power depending on the regulatory domain and frequency band of operation. TPC provides the capabilities of devices to use less than the maximum allowed output power in response to smaller path loss or larger value of the RSSI. The AP is responsible for informing stations about the maximum output power where the network is operating. The algorithm for determining the output power level is not specified in 802.11h. Note that DFS and TPC require feedback information about channel conditions. Therefore, IEEE 802.11h includes a message exchange according to which a device can be requested to measure channel conditions and report back. TPC and DFS together give the device agility and in general can be used to reduce interference and to promote coexistence with other wireless devices.

Another recent work within IEEE 802.11 is IEEE P802.11j, which will take into account Japanese regulations in the band 4.9–5 GHz and adapt IEEE 802.11a for operation in this frequency band.

IEEE 802.11g

As was discussed in the Introduction, there is always a desire to increase the data rate. The value of WLAN technology and, therefore, the user acceptance and market size depend substantially on the speed of connection. Almost as soon as IEEE 802.11b and IEEE 802.11a were approved, work began on another physical layer for the 2.4 GHz band. The requirement was at least to double the data rate of IEEE 802.11b while achieving backwards compatibility with IEEE 802.11b. Another motivation was the change in FCC regulations. While previous regulations required spread-spectrum technology in the 2.4 GHz band, the new regulations allow any digital modulation to be used.

There were two major competing technologies: an extension of PBCC and an OFDM technology, called DSSS-OFDM, which uses the preamble of IEEE 802.11b for backwards compatibility. These technologies were feverishly supported mainly by the companies that have relevant intellectual property: Texas Instruments™ (by its 1999 acquisition of Alantro Communications Inc., of Santa Rosa, CA) in the case of the PBCC and Intersil in the case of DSSS-OFDM. Neither of the technologies had qualified majority of 75% within the working group to win. As a result, the IEEE 802.11g standard is a compromise: it has one mandatory mechanism and two optional mechanisms.

The mandatory physical layer is in principal identical to 802.11a, but works in the 2.4 GHz band. 802.11g is the first standard for wireless data communications in the 2.4 GHz band which is not using spread-spectrum technology. Since 802.11a was standardized before 802.11g, and it was widely understood technology, the investment in 802.11a could be leveraged by using the same physical layer, and only changing the frequency band of operation. There are several small differences between the mandatory physical layer of 802.11g and 802.11a. The first difference is that the SIFS for 802.11a is 16 μs and the SIFS for 802.11g is 10 μs. 802.11a uses 16 μs to allow enough time for the time-consuming convolutional decoding. To give 802.11g equipment the same time to perform convolutional decoding, an 802.11g packet is followed by 6 μs of silence. The duration field of every packet includes this 6 μs signal extension. This is called a virtual extension of SIFS.

802.11g can achieve speeds of up to 54 Mb/s in the 2.4 GHz band. Recall that 802.11g is required to be backwards-compatible with 802.11b. This backwards compatibility, however, is not achieved at the physical layer. A legacy 802.11b station cannot receive any part of an 802.11g transmission. Backwards compatibility is achieved at the MAC layer. In a network composed of 802.11b and 802.11g devices, 802.11g devices can use the RTS/CTS protocol to communicate. The RTS/CTS exchange is done using 802.11b so that other stations can update their NAVs and avoid transmitting during this period. Another protection mechanism that a station can use to reserve the wireless medium for certain duration is to send CTS to itself. Everything that is mandatory for 802.11b and 802.11a is mandatory for 802.11g as well, with the following main exception: when clear channel assessment (CCA) is performed to determine whether the medium is busy, it must detect OFDM and DSSS transmissions. There are two main approaches to implement CCA: energy thresholding and detection of the 802.11b or 802.11a preambles. The only difference with 802.11b is that support for the short preamble is now mandatory.

IEEE 802.11g supports three different preamble and header formats. The long preamble is the legacy of the original IEEE 802.11. The short preamble is legacy from IEEE 802.11b. The third preamble and header format is legacy from IEEE 802.11a.

The two main contenders for 802.11g were standardized as optional mechanisms.

The first optional mechanism is DSSS-OFDM. It uses the long and short preambles defined in 802.11 and 802.11b, preceding the OFDM part. In this way, it does not need RTS/CTS or CTS-to-self protection mechanisms, because legacy 802.11b stations will be able to update their NAVs.

The other optional technology, PBCC, achieves data rates of 22 and 33 Mb/s [B49]. The two modes are called PBCC-22 and PBCC-33. PBCC-22 uses a 256-state binary convolutional code with a rate of 2/3 and a cover sequence. The input bits are divided into pairs of adjacent bits. In each pair, the first bit is fed to the upper input of the convolutional encoder and the second bit is fed to the lower input of the convolutional encoder. The output of the convolutional code is mapped to an 8-PSK constellation. This yields a throughput of two information bits per PSK symbol. The cover sequence is the same that is used in the optional PBCC mode of 802.11b. PBCC-33 mode achieves a 33 Mb/s data rate by increasing the clock frequency by 50 percent from 11 MHz to 16.5 MHz only for the data portion of the packet. (See Figure 2–32 and Figure 2–33.)

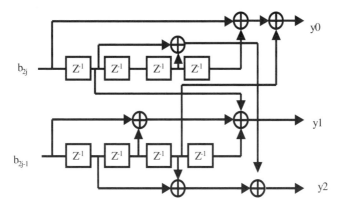

Figure 2–32:   256-state rate two-thirds binary convolutional encoder

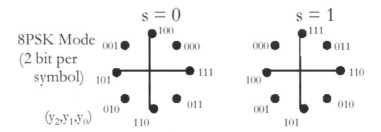

Figure 2–33:   8-PSK symbol mapping

Clearly 802.11g supports many data rates: the 6, 12, and 24 Mb/s data rates are mandatory for OFDM. In addition, the mandatory OFDM part of 802.11g has data rates of 18, 36, 48, and 54 Mb/s. The optional CCK-OFDM part of 802.11g provides the same data rates as 802.11a. The optional PBCC part of 802.11g provides data rates of 22 Mb/s and 33 Mb/s. The data rates that the 802.11 physical layers provide are summarized in Table 2–3.

Since every physical layer has multiple data rates, an important question is how to select the data rate? This is not specified in the standard and there are several approaches. Products by Lucent select the transmission rate according to how many acknowledgements the device has successfully received at a given rate. Another approach is to monitor the channel during the RTS/CTS frame exchange. It is also possible to determine the data rate on the basis of the SNR during the immediately preceding packet exchange.

Table 2–3:   802.11 physical layers

Standard	Max. Data Rate (Mbs)	Band	Frequency Range (GHz)
IEEE 802.11	2	ISM	2.4–2.485
IEEE 802.11b	11	ISM	2.4–2.485
IEEE 802.11a	54	U-NII	Low 5.15–5.25 Mid 5.25–5.35 High 5.725–5.825
IEEE 802.11g	54	ISM	2.4–2.485

802.11g supports two slot times –20 μs, which should be used when there is legacy equipment in the network, and a shorter slot time of 9 μs, which provides higher throughput and could be used when there are only 802.11g devices in the network. In an IBSS only the long slot time should be used to facilitate ad-hoc networking. The time to perform clear channel assessment should be under 15 μs if the longer slot time is used, or under 4 μs if the shorter slot time is used. The timing parameters of the three physical layers (802.11b, 802.11a, and 802.11g) are summarized in Table 2-4. The maximum contention window for all physical layers is 1023 time slots.

Table 2–4: Timing parameters for 802.11b, 802.11a, and 802.11g

Parameter	802.11b	802.11a	802.11g
SIFS duration, μs	10	16	10 (with virtual extension of 6 μs)
Slot time, μs	20	9	20 or 9
Minimum contention window, in number of slots	31	15	15

WLAN INSTALLATION

If an AP can hear another AP or a distant station, due to the very nature of the CSMA/CA protocol, it will defer transmission, just as it would for a station within its area. Similarly, if more than one AP can hear a station, all of these APs will defer transmission, thus degrading the network performance. Therefore, there is a conflict.

The WLAN installation is a significant practical problem because network performance can be significantly affected by the installation. On one hand, there should be no gaps of coverage. On the other hand, the APs should be as far apart as possible. Clearly, WLAN installation is not a trivial task. The goal of WLAN installation is to achieve certain performance with minimum cost. The performance measures are most often data rate and coverage area. In addition, there may be areas where coverage is deliberately not desired. Coexistence with other wireless networks can also be a performance measure. The WLAN installation problem involves selection of antenna types, AP locations, and AP frequencies.

First, antennas should be considered. A wide variety are available for wireless LAN systems, the most common being the dipole antenna. Most access points on the market today have dipole antennas. The dipole antenna is omnidirectional, or it radiates equally in all directions around its axis or azimuth. It does not radiate along the axis of the antenna and therefore focuses its energy in a horizontal plane with a circular pattern outward. The radiation pattern is sometimes referred to as the donut pattern. The top image in Figure 2–34 shows an example of the horizontal radiation pattern of a dipole antenna, and the lower image shows an example of the vertical radiation pattern.

If a dipole antenna is placed in the center of a single floor of a multistory building, most of its energy will be radiated along the length of that floor, with some small fraction sent to the floors above and below the access point. If the intended area is a large open area more or less symmetric in all directions, the omnidirectional antenna is certainly a good choice. In contrast to omnidirectional antennas, directional antennas concentrate their energy into a cone, also called a beam.

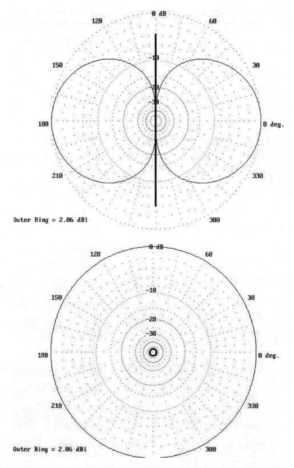

Figure 2–34: Horizontal and vertical pattern of a dipole antenna

Figure 2–35 shows the horizontal and vertical patterns of an antenna with horizontal 3-dB beamwidth of 90 degrees. A measure of the directivity is the antenna gain. While the total radiated energy is the same, directional antennas focus the energy in one direction, and therefore the gain in that direction is increased. For example, in Figure 2–35 the energy that would have been radiated to the back is redirected to the front. As a result, the antenna has a higher gain in the front and the range of the antenna in the front is greater. Note that an isotropic antenna is simply a theoretical model and it is not possible to construct one physically. An isotropic antenna has a gain equal to 0 dBi (dBi means dB in comparison with an isotropic antenna). A half-wave dipole has a gain of 2.14 dBi, or it concentrates 2.14 dB more energy in its direction of maximum radiation. In general, the gain of an antenna should take into account two other factors. The gain should be reduced by the ohmic losses in the antenna and by the mismatch factor.

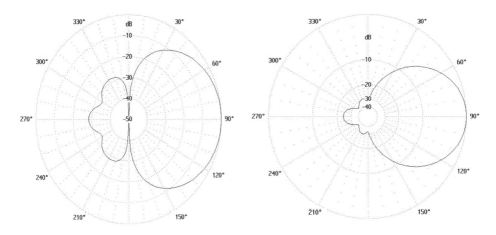

Figure 2–35: Horizontal and vertical pattern of a 90° directional antenna

Too much gain is not permitted, and regulations should always be observed. Wireless LAN access points have special connectors to keep from attaching nonpermissible antennas. Modifying the connector to accept higher gain antennas is not permitted. It is allowable to use higher gains than 6 dBi, provided the transmitter power is reduced by the same amount that the antenna exceeds 6 dBi.

This means that observing the gain and antenna pattern together is needed when choosing directional antennas. An antenna such as this could be placed in the corner of a work area, concentrating the energy into the work area and radiating very little outside the building. Directional antennas are normally available in patterns of 30, 45, 60, 80, 90, 120, and 180 degrees. In addition, some antennas have adjustable beamwidths, but are more expensive.

Another antenna issue is antenna diversity. Antenna diversity is simply taking the uncorrelated signals from two or more different antennas and combining them for added signal strength. Cellular and PCS systems have been using two receive antennas at the base station to reduce multipath fading for many years. It is now becoming common to use diversity in wireless LAN systems. There are several types of diversity: spatial, polarization, pattern (angle), temporal (time), and frequency diversity. Only spatial, polarization, and pattern diversity are practical for wireless LAN antenna systems. Currently, spatial diversity is the most widely used form of diversity. Some access point products have two antennas, such as Proxim's Harmony. One antenna is used as the primary transmitting and receiving port, while the other is periodically polled to see if it is receiving a stronger signal than the main antenna. It is known that signals received from diversity antennas spaced half a wavelength apart can be essentially uncorrelated, depending on the channel impulse response. In the 2.4 GHz range, half a wavelength is 62 mm, or 2.5 in., and in the 5.8 GHz range, half a wavelength is 26 mm, or about 1 in. Therefore, spatial diversity can be employed in consumer products.

Polarization diversity is the placement of antennas or antenna elements separated by an angle of 90 degrees. When a signal hits an object, such as a metal cabinet, a wall, etc., not only is there reflection, but there is also polarization change. Positioning two antennas with different polarizations can increase signal strength by up to 3 dB.

Another type of diversity that also makes sense is pattern or angle diversity. It involves using two or more different radiation patterns combined to form a more suitable pattern. Pattern diversity can be used in PC cards to make the radiation pattern for the laptop close to a hemisphere above the laptop. This radiation pattern would maximize robustness and coverage and can be achieved with a dipole antenna with a directional antenna (Figure 2–36).

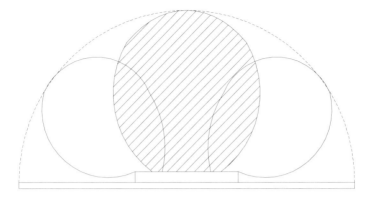

Figure 2–36: Pattern diversity on the laptop.

The use of smart antennas [B104] with IEEE 802.11 WLANs has not been extensively considered. It is not impossible, but at present, the standard does not efficiently support smart antennas [B115]. While in the past, the cost of smart antennas has been high for consumer products, the price has been decreasing. In the future, perhaps they can be used by WLANs for increased performance. At present, some companies offer smart antenna solutions for 802.11. The main direct advantage of smart antennas is reduced transmit power. This makes frequency reuse possible, improving the capacity and battery life. One problem with smart antennas is that the position of the antenna to achieve the correct beam pattern is not known in advance. A possible solution is to send the RTS and CTS packets with omnidirectional pattern and obtain channel information during the RTS/CTS exchange [B112]. During the RTS/CTS exchange, stations determine the appropriate beamforming, which will be used during the data exchange. This still allows a great deal of flexibility and, therefore, is an opportunity for vendor differentiation. Smart antennas have been used together with PCF operation [B115].

> Note that the use of smart antennas makes the support of mobility more difficult, because even a small movement will have a significant effect. To support mobility with smart antennas, advanced adaptive or blind signal processing steps are needed.

To determine the AP and antenna placement, two approaches can be used. The first approach is to perform a site survey. The second approach is to use computer simulations. A site survey is done by performing direct RF measurements. Some prior planning is needed before the actual site survey is performed regarding usage overlap and number of users per AP. A 20 percent overlap in coverage area is recommended, because laptops on the edge may move in and out of the coverage area either by physical movement or the by fading of the access point's signal. The overlap in coverage creates a problem, however. The performance of the network will depend on the algorithm that the station uses to select an AP. An algorithm that selects the AP with the strongest signal will not always provide best overall results. Other criteria that could be used include capabilities and channel conditions. To facilitate AP selection and roaming, it is desireable for stations to receive information about channel conditions from the APs. This is not supported by the standard at present and is being investigated by 802.11k, a Task Group devoted to radio resource management. For example, the APs could include information about channel conditions and neighboring APs in their beacons.

In the site survey, floor plans and building material need to be considered. Concrete and steel attenuate the signal more than other building materials. RF interferers can be identified with spectrum analyzers. As a rule of thumb, APs should be kept at least 20 feet from microwave ovens or other access points.

An example of a floor plan is shown in Figure 2–37. Note that the main obstacles to RF transmission are the elevator shafts, marked with large Xs.

The floor plan shown in Figure 2–37 can be covered with some overlap, as shown in Figure 2–38.

The goal of installing a WLAN is to maximize coverage with as few access points as possible. It is difficult to use computer simulations. For example, floor plans may not be available in suitable electronic form, and it is difficult to take everything into the computer model. A site survey, together with directional and diversity antennas, will yield the best results in practice. Many offices have narrow and odd-shaped areas. Narrow beamwidth antennas can effectively cover long hallways, wider beamwidth antennas are more suitable in wide areas, and 90 degree antennas are a good choice for placement in corners. Furthermore, large spaces have large delay spreads, making diversity antennas effective. In the floor plan in Figure 2–37, the use of directional antennas allows the area to be covered with four antennas instead of six (Figure 2–39) [B41].

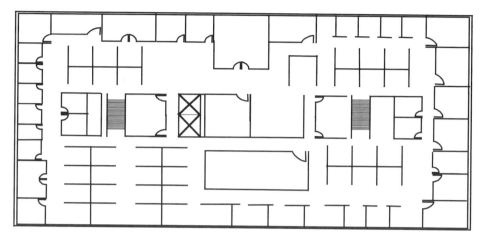

Figure 2–37:   Typical floor plan

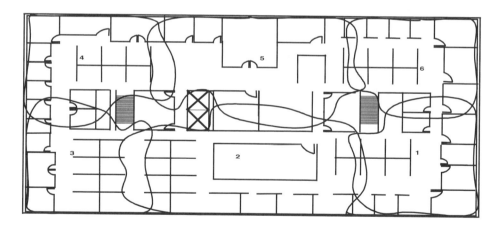

Figure 2–38:   An example of an WLAN installation

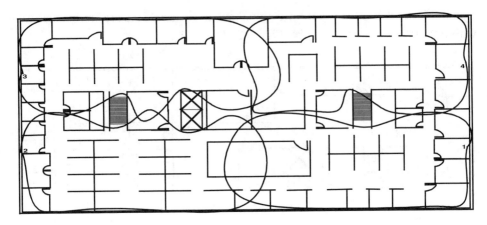

Figure 2-39: An installation of WLAN with directional antennas

After the APs have been located and their coverage areas measured, radio channels are assigned to the APs. A total of eleven DSSS radio channels are available in the 2.4 GHz band, and of these, three have minimal spectral overlap, as shown in Figure 2-23 and Figure 2-24. It is interesting to note that making these frequency assignments is essentially a map-coloring problem with three colors, and there are various algorithms that give optimal or near-optimal assignment of the three radio channels, given the AP locations and coverage areas.

Another factor to consider in the design of a WLAN is areas with high and low densities of users. Areas of high densities are convention centers, airports, etc. Obviously, high-density areas will need more APs. The number of necessary APs N can be determined in the following way:

$$N = \frac{Th_{user} \times N_{users} \times A}{\text{Data rate} \times \text{efficiency}} \qquad \text{Eq. 2-11}$$

where Th_{user} is the desired throughput per user, N_{users} is the number of users, A is the average activity rate per user, Data rate is the data rate at which users communicate with the APs, and efficiency is the throughput efficiency in percentage terms at this data rate. For example, for an 802.11a network, suppose that there are 100 users, and each user communicates with the APs at 24 Mb/s with efficiency equal to 0.6 (throughput of 60 percent), and each user is active only 20 percent of the time. If the required throughput per user is 3 Mb/s, the number of APs necessary to provide the required throughput is

$$N = \frac{3 \times 100 \times 0.2}{24 \times 0.6} \qquad \text{Eq. 2-12}$$

Since the number of APs is an integer, in this example, 5 APs must be used to provide the desired performance.

It may be desirable to reduce the output power of some APs, because this will reduce their coverage areas. As a result, they will serve fewer users. To achieve this, some WLAN products allow one to set receiver threshold, thus controlling the size of the coverage area. Normally the receiver threshold will be set to a minimum to maximize the coverage area. In high-density areas, the threshold may be set to a higher value to reduce the coverage area.

Another factor to help select an appropriate AP is the QoS that can be expected. Therefore, as part of 802.11e APs will broadcast in the beacons information on the number of stations that are associated with them, channel utilization, and the remaining amount of bandwidth capacity available. Channel utilization is defined as percentage of time the AP has determined the medium to be busy. Therefore, roaming stations can select an AP that is likely to accept future admission control requests.

IEEE 802.11 TECHNOLOGY AND BUSINESS TRENDS

The market success of 802.11 and widespread proliferation of devices has caused some additional technical problems. As was said in earlier, equipment must be certified by the regulatory body in the country of sale. In addition, most radio regulations do not permit transmissions by equipment that is not certified for those regulations. Wireless devices, however, do not understand in which country they are located. The 802.11d part of the standard tries to deal with this issue. Stations that implement IEEE 802.1d will maintain a list of countries in which they are certified to operate and the transmission parameters for these countries. These stations can obtain information about the regulatory domain in which they are located, without sending any transmissions. This regulatory domain information (such as the country codes defined in ISO/IEC 3166-1:1997) is in the beacon frame and stations will use only passive scanning initially. Active scanning is not allowed at first. Only after obtaining the regulatory domain information, can the station determine if it is certified to operate in this regulatory domain. Only after this determination is done, can a station legally transmit and perform, for example, active scanning. 802.11d is not specific to any particular physical layer.

Yet another 802.11 project that was launched in Nov. 2002 is 802.11k for radio resource management. The goal is to provide interfaces for radio and network measurements. Radio resource management provides mechanisms that support intelligent admission of sessions, and distributions of traffic and power. The goal of radio resource management is to achieve optimized usage of radio resources and maximum system capacity. The current draft of P802.11k [B76] is still in an early stage of development. According to the draft, there are dedicated and concurrent measurements. Dedicated measurements require the interruption of the normal operation of a device. Concurrent measurements are performed during the course of normal operation. Measurement start times are random. 802.11k will also try to identify hidden nodes and produce hidden node reports.

Motivated by the need for higher data rates, 802.11 established Task Group n to develop a standard for an even higher data rate systems. One simple method to increase data rate is to use a higher-order modulation, for example 256 QAM, or 512 QAM. This, however is not efficient, because these modulation schemes require very high signal-to-noise ratios and the data rate will be increased only at very small distances, about 10 meters. Another technique to increase the data rate of 802.11a (and 802.11g) is to reduce the sampling interval from 50 *ns* to 25 *ns* to and preserve the same number of carriers—64. This will have the effect of reducing the symbol period by a factor of 2, from 4 μs to 2 μs. However, the time duration of the cyclic prefix is also reduced by 2 and becomes 400 *ns*. If the delay spread of the channel is longer than 400 *ns*, this approach will lead to increased ISI, if additional equalization is not used. Another simple approach is to double the number of subcarriers from 64 to 128. This also reduces the sampling interval to 25 *ns*, but preserves the symbol period and the duration of the cyclic prefix. These approaches perform a little better compared with the simple use of a higher-order modulation.

Another approach to increase the data rate is based on the following idea. IEEE 802.11a uses the same QAM constellation on all 48 data carriers. This encodes the same number of bits on all the carriers. A better approach is to use a different

QAM constellation for every carrier, i.e., to encode different number of bits per carrier [B25], [B100]. In a multicarrier system, the process of modulating different number of bits on each carrier based on the SNR of the carriers is often called bit loading. Bit loading is much better suited to the characteristics of a multipath channel. Bit loading also enables constellations sizes much higher than 64QAM to be used. One design requirement in a system that uses bit loading is that it requires a feedback mechanism to communicate the bit-allocation table from the receiver to the transmitter. This communication from the receiver to the transmitter must happen faster than substantial changes in the channel profile.

Further improvement in data rate can be obtained combining bit loading with coded modulation techniques (such as trellis coded modulation (TCM)). TCM maps the coded bits to points in the signal space such that the Euclidean distance between transmitted sequences is maximized. Typically the method of set partitioning is used to maximize the Euclidean distance. Since coded modulation schemes do not encode all information bits, coded modulation must be combined with bit loading in multipath channels to achieve the coding gain benefits. If the average number of bits per carrier remains the same, this technique will not lead to a higher data rate. Bit loading together with trellis coding allows for a higher average number of bits per carrier. Figure 2–40 [B25] illustrates the higher data rate that can be obtained with the new approach for the lower UNII band. The results on this figure are obtained assuming transmit power of –56.9 dBm/Hz (this is less than 40 mW), receiver noise figure of 10 dB, and an additional implementation loss of 5 dB. The thermal noise level is assumed to be –174 dBm/Hz. The gain of the receive and transmit antennas is assumed to be zero.

In Figure 2–40, the 802.11a system uses equal number of bits per carrier. The maximum data rate for the two systems is determined as the maximum data rate in which the packet error-rate is 10 percent. Since in the simulation the maximum output power is considerably less than the maximum per FCC regulations significant further improvement in range can be obtained. Yet another approach for high-data rate wireless systems is based on a technology called vector OFDM, or multiple-input multiple-output OFDM (MIMO OFDM) [B8], [B94], [B149], [B150]. This technology achieves considerably higher capacity, but requires computationally intensive signal processing operations. Finally, note that it is possible to combine several or all of the above-mentioned methods to achieve even higher data rates. Some early research indicates that it is possible to achieve data rates in excess of 250 Mb/s [B43].

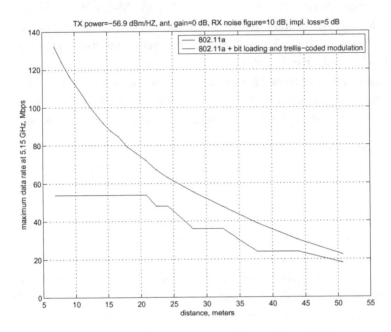

Figure 2–40: Comparison of the maximum data rates that can be achieved with the specified simulation parameters for a plain 802.11a system and 802.11a system with bit loading and trellis coding

The 802.11n Task Group realized from the beginning that throughput must be increased, and not just the maximum data rate. Throughput, however, depends primarily on the MAC protocol [B145]. Therefore, 802.11n will likely develop not only a physical layer, but also appropriate MAC modifications to ensure high throughput.

In July of 2003, within IEEE 802.11, a Study Group on Dedicated Short-Range Communications (DSRC) in the 5.9 GHz band was founded. As in the past, the direct motivation was the allocation in 1999 by the FCC of 75 MHz of bandwidth (5.85–5.925 GHz) right above the higher UNII band for communication between vehicles and roadside systems and between vehicles. Similar bands exist in Japan and Europe. In addition, further FCC rules were expected by the end of 2003.

Furthermore the U.S. Congress directed the Department of Transportation to develop a standard to ensure interoperability in the DSRC band. This resulted in the ASTM E2213 standard [B128], developed by the American Society for Testing and Materials (ASTM), and largely based on 802.11a. The standard is written as an amendment, specifically allowing 802.11a chipsets to be used in DSRC equipment. The MAC modifications introduced by ASTM E2213 can be satisfied using a simple firmware upgrade. The applications of this standard are numerous and include safety enhancements (intersection collision warning, etc.), roadside kiosks (updated map information, weather updates, road conditions), Internet access hot spots at service stations and parking lots, electronic toll payment, inventory tracking, etc. The 802.11 Study Group was established because the ASTM finds it more appropriate for the DSRC standard to be developed further and maintained by 802.11. As 802.11 evolves, it is easier to update DSRC together with 802.11.

The traditional view of wireless communication networks is that there are two network categories: voice-oriented and data-oriented. Each of these categories can be further subdivided into two subcategories depending on the intended coverage area: local area or wide area.

Wireless voice systems were developed long before wireless data systems. Wireless voice systems have been in use in the USA. since the 1930s when the first police vehicles were equipped with one-way and then two-way radios. Cellular telephony was technically developed at Bell Telephone Laboratories in the 1940s and commercially deployed on a wide scale in the 1980s. Simultaneously with the development of 802.11, cellular telephony has made enormous progress in its ability to handle not only voice, but also data. Third-generation cellular telephony will offer up to 2 Mb/s at a fixed location, although the real throughput will be much lower. In the meantime, before third-generation networks are widely deployed, the so-called 2.5 generation of technology will offer data rates up to 172 Kb/s, with actual speeds of about 42 Kb/s.

The main business trend in cellular telephony is the market saturation; slow growth in revenue from voice services and higher growth in revenue from data services. Clearly the fundamental long-term trend is the merger between voice-only and data-only networks. This merger of voice and data networks requires seamless roaming between cellular and wireless local area networks [B101], [B117]. Already there are technologies and products on the market that enable single-subscriber identity on these two different networks. Future wireless networks will be a combination of several radio communication technologies, such as third generation radio access and wireless LAN [B12]. With top speeds of 54 Mb/s WLAN technology will continue to provide much superior bandwidth compared to any cellular technology.

At present commercially available WLAN devices are predominantly network adapters (NAs) and access points (APs). NAs are available as PC cards, physically separate from laptops. The market though is moving towards integrated devices. Many laptops sold at present have integrated WLAN interfaces with 802.11b, and some with 802.11a and 802.11g. In the next several years WLAN devices will be integrated in many consumer electronic products as well.

Today most large offices are equipped with appropriate wiring for conventional LANs. The first market for wireless LANs is large indoor areas and offices and buildings with wiring difficulties, such as historic buildings, etc., large indoor areas such as manufacturing floors, warehouses, convention and exhibition centers, libraries, etc. WLAN technology is increasingly the network of choice. In today's fast-paced world with frequent relocations of equipment and high number of temporary offices, portability significantly increases productivity. A major factor in the market growth of WLANs is that the market for laptops and other portable computing devices is growing faster than the desktop market. Obviously portable computing devices naturally call for portable LAN, i.e., WLAN. WLANs also will create new markets for equipment and services.

Overall, 802.11 not only improves business productivity, but is changing the dynamics of some business. For example, at conferences 802.11 allows instant polling, instant journalism, etc.

An interesting business question to consider is to what extent will WLAN compete and to what extent will it complement other wireless data communication technologies. There is no simple answer to this question—it depends on the business model. To some cellular service providers offering high-data rate service 802.11 may appear as a foe. Clearly combining 802.11 with directional antennas for extended coverage creates more significant competition to wide area wireless networks. Many companies have started providing high-speed data services using

WLANs in places such as coffee shops, fast food restaurants, airports, convention centers, etc. T-Mobile is the largest WLAN service provider at present. A number of start-up companies have business models involving providing WLAN service.

IEEE Std 802.11 is a mature and low-cost technology, and in addition operates in the unlicensed spectrum. At the same time mobile phone service providers are introducing 3G technology in the marketplace. The cost per customer of 3G networks is higher than that of WLAN because of higher equipment costs and operation in expensive spectrum. In addition, WLAN offer higher data rates compared with 3G networks. Table 2–5 shows a comparison between cellular telephony and 802.11.

Table 2–5: Comparison between cellular telephony and 802.11

Parameter	Cellular	802.11
Range	High	Low
Coverage	Ubiquitous	Localized to "hot spots"
Power consumption	Low	Medium
Spectrum	Licensed (very expensive)	Unlicensed
Security	Very secure	WEP is broken, but will be secure with IEEE Std 802.11i
Data rate	Low	High
Infrastructure cost	Very expensive	Very cheap

Wireless local area networks are used to provide data-only service. 802.11 is not used to provide voice service and this makes it more suitable as a complement to 3G systems. There are several market opportunities with regard to WLAN and 3G. First, since users value highly seamless mobility and service, some products enable seamless transition between WLAN and 3G networks. When WLAN coverage is available, stations can communicate at a higher data rate. In the absence of WLAN coverage, stations can communicate over 3G networks. The two networks complement themselves: one with wide coverage, and the other with enhanced performance in isolated areas. A more interesting question is whether a complete outdoor cellular network can be designed, based on 802.11 [B22]. Such a network would have at the same time cost and performance advantages over current 2G, 2.5G, and 3G networks. The main problem of an 802.11-based cellular network is the limited transmission range. This problem was recently investigated by Clark, et al., in their article, "Outdoor IEEE 802.11 cellular networks: radio link performance" [B22]. The main conclusion is that yes, an outdoor cellular network based on WLAN is feasible, with a cell radius in the range of 0.7 to 3 km, assuming 802.11b and 1 Mb/s data rate. However, this requires additional signal processing, since the rms delay spread for outdoor urban areas is about 1 microsecond. To mitigate this large delay spread a RAKE receiver is necessary.

While the first big wave in the deployment of 802.11 was its use in offices, the next big wave is expected to be the "unwiring" of consumer electronics devices such as TV sets, CD and DVD players, etc. Since this will unleash the long-awaited demand for video over the Internet, this will likely have a significant and much-needed impact on the whole communications industry. The acceptance of 802.11a by the computer industry will likely push 802.11a as a link between consumer electronic and computer devices.

Not only is the number of Wi-Fi hot-spots growing, but there are places where these hot spots are adjacent and installed in a way to provide continuous coverage. Cities with the most installed access points are San Francisco, followed by New York, Seattle, Chicago, and San Jose. This does not include accidental hot spots leaking out of private homes and offices. In the years ahead 802.11 will become a truly universal standard, found everywhere in the world. Current market research estimates that 99 million people will be using 802.11 by the year of 2006.

Chapter 3 Standards for Wireless Personal Area Networking (WPAN)

INTRODUCTION

While the growth and success of WLAN is due to the market growth of laptops, the main market demand for wireless personal area networking (WPAN) technology comes from the fast growth of small and inexpensive devices many of which have Internet connectivity. Examples of these electronic devices are cellular phones, personal digital assistants, personal music players, digital cameras, etc. As was mentioned in Chapter 1, it is a basic law of communications technology that the value that these devices represent to consumers would be significantly enhanced if they were able to communicate among themselves and with other devices. Traditionally, proprietary special-purpose cables have been used to solve this short-range connectivity problem. However, most users find that cables not only limit mobility, but are outright frustrating. Therefore, it becomes quite desirable to develop connectivity technologies for interconnecting personal devices that do not require the use of cables. These connectivity technologies are known as wireless personal area networks (WPANs).

COMPARING WPAN AND WLAN

What are the similarities and differences between WPAN and WLAN? At a first glance, a WPAN may appear to resemble a WLAN, like IEEE 802.11. Both the WLAN and WPAN are short-range wireless data communications technologies. However, WLANs have been designed and are optimized for portable computing client devices such as notebook computers. WPAN devices are even more mobile. The two technologies differ in three ways: power levels and coverage, control of the media, and lifespan of the network.

Typical WLAN coverage distances are on the order of 100 m with transmit power of 100 mW, and about 500 m with transmit power of 1 W. Furthermore, power consumed by the device can be significant. Mobile devices typically operate on batteries. Portable devices are normally used while in a fixed location and therefore usually will run from power supplied by wall sockets. WLANs are typically wireless extensions of wired LANs. The devices participating in a WLAN are mainly of one type (laptops), while the devices participating in a WPAN can be of several different types. WPAN devices cover areas of about 10 m with transmitted power of about 1 mW. WPAN technology includes additional measures to ensure low power consumption by the device, making battery operation possible for long enough periods.

 Note the distinction between "mobile" and "portable" devices. Mobile devices typically operate on batteries. Portable devices are normally used while in a fixed location and therefore usually will run from power supplied by wall sockets.

Because WLANs cover relatively large areas with many uncoordinated devices, the mechanism for access to the medium must handle potential collisions. CSMA/CA, used in IEEE 802.11, is a very efficient protocol for WLANs. WPANs cover small areas where devices can be synchronized, significantly reducing collisions. As a result, the medium access protocol can be different from the one used for WLANs.

The final difference between WPAN and WLAN is the different lifespan of the networks. As bona fide members of a larger infrastructure, WLANs do not have inherent or implied lifespan. They have "existence" independent of their constituent devices. This is not true for WPANs. WPAN networks are ad-hoc, i.e., they are created in a spontaneous fashion, cover a small area, and exist for a small period of time. Connection establishment for ad-hoc networks must be very fast. Personal devices that participate in a WPAN are designed for their personal appeal and functionality. They are not designed to be members of an established networking infrastructure, although they may connect to it when necessary. In a WPAN, a device creates a connection that lasts only for as long as needed and has a finite lifespan. For example, a file transfer application may cause a connection to be created only long enough to accomplish its goal. When the application terminates, the connection between the two devices may be severed as well. The devices to which one's personal device is currently connected in a WPAN may bear no semblance to the device that it was previously connected to or it will connect to next. For example, a notebook computer may connect with a PDA at one moment, a digital camera at another moment, or a cellular phone at yet another moment. At times, the notebook computer may be connected with any or all of these other devices.

The demands of the consumer market require WPAN technology that is inexpensive, consumes very little power, and efficiently supports QoS. Further, these devices will likely be used in many different environments such as offices, homes, convention centers, shopping malls, airports. Therefore, these devices must satisfy the marketing requirements dictated not only by the consumer market but the business market as well.

The range of operation is short, but there may not be a line of sight. The physical layer must work in a non-line-of-sight (NLOS) environment.

The IEEE 802.15 Working Group has taken a leading role in WPAN technology development. First, the Bluetooth protocol [B13] was standardized as IEEE Std 802.15.1-2002 [B78]. To achieve this, the IEEE 802 Working Groups received a nonexclusive license from the Bluetooth Special Interest Group. However, IEEE 802.15.1 not only standardized Bluetooth, but in the process of developing IEEE 802.15.1 on the basis of Bluetooth, IEEE 802.15.1 made significant contributions to the Bluetooth specification. Second, the 802.15 Working Group undertook serious work to develop a recommended practice for the coexistence of IEEE 802.11b and Bluetooth. The results of this effort became IEEE Std 802.15.2™-2003 [B79]. Third, IEEE 802.15 developed two other WPAN technologies—a high-rate WPAN (IEEE Std 802.15.3™-2003) [B80] and a low-rate WPAN (IEEE Std 802.15.4™-2003) [B81]. This chapter is devoted to these technologies.

The organization of the chapter follows the standardization work within IEEE 802.15. First, IEEE 802.15.1 is presented. The coexistence mechanism is discussed next, followed by a discussion on the high-rate WPAN. And finally, the low-rate WPAN is presented.

IEEE 802.15.1

Overview

The first WPAN standard is Bluetooth, developed by an industry consortium called Bluetooth Special Interest Group (SIG). The original intent was that Bluetooth would be like a cable replacement with a range of operation of about 10 m, very little power consumption, and very inexpensive. Thus Bluetooth is

Chapter 3: Standards for Wireless Personal Area Networking (WPAN)

optimized for power-conscious, battery-operated, small size, lightweight personal devices.

> Led initially by Ericsson®, Nokia®, IBM®, Intel®, Toshiba® and later joined by 3Com, Motorola®, Lucent, and Microsoft®, the Bluetooth SIG has about 2000 member companies, which demonstrates the enormous interest in the industry [B53].

A Bluetooth WPAN supports both synchronous communication channels for toll-quality voice communication and asynchronous communications channels for data communications. Unlike IEEE 802, the Bluetooth standard covers all layers in the OSI model. However, an IEEE 802-style MAC and PHY are not used. The precise mapping between the elements of the Bluetooth protocol and IEEE 802-style MAC and PHY is cumbersome and debatable. It was attempted but ultimately abandoned by the IEEE 802.15 Working Group. Figure 3–1 shows the protocol stacks in the OSI 7-layer model and the Bluetooth wireless technology and their relation as it pertains to this standard. As shown in Figure 3–1, the logical link control (LLC) and MAC sublayers together encompass the functions intended for the data link layer of the OSI model.

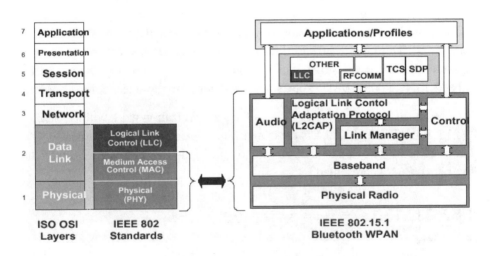

Figure 3–1: The mapping between the Bluetooth protocol and the IEEE 802 architecture

136 *Wireless Communication Standards*

IEEE 802.15.1 is derived from the Bluetooth version 1.1 foundation core [B13]. To make Bluetooth an IEEE 802 standard, the Bluetooth SIG granted the IEEE a nonexclusive license to copy and adapt the Bluetooth specification.

To ensure the commercial success of Bluetooth and enable the creation of interoperable, and interactive applications, the Bluetooth specification in addition to the communication protocols includes interoperable applications. These applications support various usage scenarios addressed in the specification. IEEE 802.15.1 covers only a subset of the communications protocols in the Bluetooth specification related to PHY and MAC protocols as identified in Figure 3–1. It includes both Bluetooth-specific protocols such as the link-manager protocol (LMP) and the logical link control and adaptation protocol (L2CAP) and non-Bluetooth-specific protocols, which are grouped in the "Other" box in Figure 3–1.

The RFCOMM layer is a serial port emulation layer for enabling legacy applications over Bluetooth links. The TCS is a telephony control and signaling layer for advanced telephony applications. The SDP is a service discovery protocol layer allowing Bluetooth devices to inquire other devices for the services that they can provide. These layers are not part of IEEE 802.15; they are specified only in the Bluetooth specification.

The L2CAP protocol supports higher-level protocol multiplexing, and the requirement for this multiplexing segmentation and reassembly. Due to the small size of the packets, large upper-layer packets need to be segmented before transmission over the air.

The Bluetooth specification refers to the control protocol, (the rightmost block in Figure 3–1), as the host-control interface (HCI). It has two functions. First, as shown in the figure, it provides physical interface to higher layers. Because IEEE 802 standards do not specify implementation, this type of interface is outside the scope of IEEE 802 protocols. Second, it provides access to hardware status and control registers. This function is implemented by the management sublayer in IEEE 802 standards.

IEEE 802.15.1 Physical layer

The Bluetooth WPAN operates in the unlicensed 2.4 GHz, the Industrial, Scientific, Medical (ISM) band. The center frequency of these channels for the

United States, Europe, and most of the world is $f_c = 2402 + k$ MHz, where $k = 0,\ldots,78$. France has different regulations in the 2.4 GHz band, and the center frequency is $f_c = 2454 + k$ MHz, where k takes values between 0 and 22.

According to the output power, the equipment is divided into three power classes. Class 1 equipment has a maximum transmit power of +20 dBm (100 mW) and minimum output power of 0 dBm (1 mW). These power level requirements are, like in all wireless 802 standards, at the antenna connector of the equipment. The output power of class 1 devices can be adjusted in steps ranging between 2 dB and 8 dB between the maximum and minimum power levels. This adjustment of power levels in a monotonic sequence is called power control and is required for class 1 equipment. Power control achieves not only power efficiency, but minimizes potential interference to other wireless systems.

Note that the transmitter will not know how to adjust its output power level without feedback from the receiver. Therefore, power control can be used only if the receiver can provide feedback; i.e., the receiver can perform measurements (such as received signal strength indicator [RSSI] levels) and there is a protocol to exchange information with the transmitter about the power levels. Clearly, power control cannot be used before a connection is set up. At connection set up the devices exchange information about their capabilities. If the receiver is not capable of providing appropriate feedback, then the transmitting device must switch from class 1 to class 2 or class 3.

Class 2 equipment has a maximum transmit output power of 2.5 mW (4 dBm) and a minimum output power of 0.25 mW (–6 dBm). Class 3 equipment has a maximum output power of 1 mW (0 dBm) and no minimum. Power control capability for class 2 and class 3 devices can be used, but is optional.

A fast frequency hopping (1600 hops/sec) transceiver is used to combat interference and fading in this band. A slotted channel is used, which has a slot-duration of 1/1600=625 µs. Each packet is transmitted on a different frequency in the hopping sequence. Channel spacing is 1 MHz. High-power efficiency and low-transceiver cost cannot be achieved with modulation schemes like quadrature amplitude modulation (QAM). Bluetooth uses a constant-envelope modulation called Gaussian Frequency Shift Keying (GFSK). A binary 1 is represented by a positive frequency deviation, and a binary 0 is represented by a negative frequency deviation. The transmitted data has a symbol rate of 1 Msymbol/s.

Baseband

Network architecture

The basic network is called a piconet. A frequency-hopping channel based on the address of the master defines each piconet. All devices participating in communications in a given piconet are synchronized to the frequency-hopping channel for the piconet using the clock of the master of the piconet. Slaves communicate only with their master in a point-to-point fashion, under the control of the master. The master's transmissions may either be point-to-point or point-to-multipoint.

 A piconet is a WPAN formed by a Bluetooth device serving as a master and one or more Bluetooth devices serving as slaves. Up to seven slaves can be active in the piconet. In addition, up to 255 slaves can remain locked to the master in a so-called parked state. These parked slaves cannot be active on the channel, but remain synchronized to the master.

Note that the names "master" and "slave" only refer to the protocol—all Bluetooth units are identical. Although certain applications may require that some devices act always as masters or slaves, this situation is outside the scope of the 802.15.1 standard. A slave device at one communications session could be a master in another and vice versa.

A scatternet is a collection of piconets, i.e., a scatternet is a collection of piconets that overlap. A Bluetooth device may participate in several piconets at the same time. Each piconet can only have a single master. Slaves can participate in different piconets on a time-division multiplex basis. In addition, a master in one piconet can be a slave in another piconet. The piconets are not frequency-synchronized. Because each piconet has a different master, the piconets hop independently, each with their own channel-hopping sequence and phase as determined by the respective master. In addition, the packets carried on the channels are preceded by different channel access codes as determined by the master device addresses. (The manufacturer assigns each Bluetooth device a unique 48-bit Bluetooth device address.) As more piconets are added, the probability of collisions increases—a graceful degradation of performance results, as is common in frequency-hopping spread-spectrum systems.

If multiple piconets cover the same area, a unit can participate in two or more overlaying piconets by applying time multiplexing. To participate on the proper channel, it should use the associated master device address and proper clock offset to obtain the correct phase. A Bluetooth unit can act as a slave in several piconets, but only as a master in a single piconet. Because two piconets with the same master are synchronized and use the same hopping sequence, they are one and the same piconet.

The scatternet topology poses the question of efficient routing, which is not answered in the 802.15.1 standard. One suitable routing protocol has been developed in Prabhu and Chockalingam's paper, "A routing protocol and energy efficient techniques in Bluetooth scatternets" [B120]. The protocol in [B120] uses the available battery power in the devices as a cost metric. This routing protocol exploits the master-slave switch and the power control capability to achieve near optimal performance with respect to battery life. The master may drain its batter faster than the slaves; therefore, the master-slave switch can be performed (depending on the remaining battery power) and can increase network lifetime.

From the architecture of IEEE 802.15.1, it is clear that it provides a point-to-point connection (master-to-slave or slave-to-master) or a point-to-multipoint connection (broadcast messages from the master to all the slaves). The topology of a Bluetooth network is shown in Figure 3–2. By definition, the master is represented by the Bluetooth unit that initiates the connection to one or more slave units.

Every Bluetooth unit has an internal system clock that determines the timing and hopping of the transceiver. The Bluetooth clock is derived from a free, always running native clock. These native clocks obviously are not synchronized. Synchronization is very important for the Bluetooth standard, because the medium access mechanism follows TDMA principles. To achieve synchronization with other units, offsets are used that are added to the native clock. The timing and the frequency hopping on the channel of a piconet is determined by the Bluetooth clock of the master. After the piconet is established, the master clock is communicated to the slaves. Each slave adds an offset to its native clock to be synchronized to the master clock. Because the clocks are free-running, the offsets have to be updated regularly. The clock determines critical periods and triggers the events in the Bluetooth receiver.

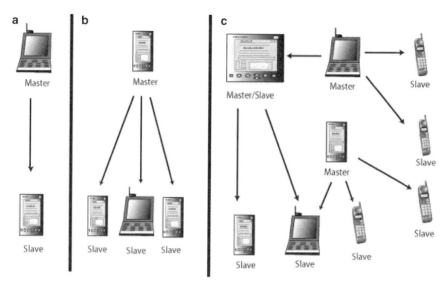

Figure 3–2: Topology of a Bluetooth network: (a) point-to-point, (b) point-to-multipoint, and (c) scatternet

Channel access and QoS

The channel is divided into time slots, each 625 μs long. The time slots are numbered according to the Bluetooth clock of the piconet master. The slot numbering ranges from 0 to $2^{27} - 1$ and is periodic. The channel is represented by a pseudorandom hopping sequence, which hops through the 79 or 23 RF channels. The hopping sequence is unique for the piconet and is determined by the Bluetooth device address of the master. The Bluetooth clock of the master determines the phase in the hopping sequence.

On the channel, information is exchanged through packets. A packet nominally covers a single slot, but it can be extended up to either three or five slots, as shown in Figure 3–3. The number of slots must be an odd number. Each packet is transmitted on a different hop frequency. The RF frequency is fixed for the duration of the packet. The nominal hop rate is 1600 hops/s.

Chapter 3: Standards for Wireless Personal Area Networking (WPAN)

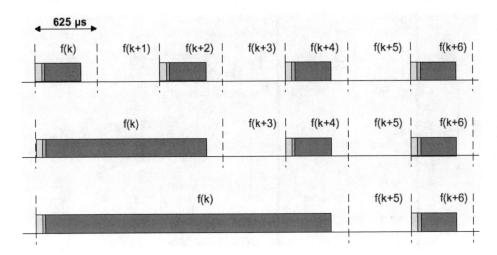

Figure 3-3: One-, three-, and five-slot packets

Each packet consists of three entities: the access code, the header, and the payload. The access code and header are of fixed size: 72 bits and 54 bits, respectively. The payload can range from zero to a maximum of 2745 bits, depending on the packet type. Packets may consist of the (shortened) access code only, of the access code-header, or of the access code-header-payload. The header has information about the packet type and to whom it is addressed. The access code is used for detection, synchronization, DC offset compensation, and piconet and device identification. The access code identifies all packets exchanged on the channel of the piconet—all packets sent in the same piconet are preceded by the same channel access code. The access code is also used by the receiver to perform robust detection, timing synchronization, and offset compensation. For example, detection and synchronization can be performed by sliding correlation of the received signal against the access code. If sliding correlation is computationally complex, other methods can also be used. The access code is also used in paging and inquiry procedures. In this case, the access code itself is used as a message, and neither a header nor a payload is present.

There are three error correction schemes defined for Bluetooth: 1/3 rate forward error correction (FEC) and 2/3 rate FEC, as well as an automatic repeat request (ARQ) scheme for the data. The purpose of the FEC scheme on the data payload is to reduce the number of retransmissions. However, in a reasonably error-free environment, FEC gives unnecessary overhead that reduces the throughput. Therefore, while packet headers are always protected by a 1/3 rate FEC, the payload is not always protected. The 1/3 FEC is a simple three-times repetition code. The repetition code is implemented by repeating the bit three times. The other FEC scheme is a (15,10) shortened Hamming code. This code can correct all single errors and detect all double errors in each codeword. The third FEC approach is ARQ.

Before the user information is sent over the air interface, several bit manipulations are performed in the transmitter to increase reliability and security. To the packet header, an HEC is added, both the header and the payload bits are scrambled, and FEC coding is applied (Figure 3–4 and Figure 3–5). Both the header and the payload are scrambled with a data-whitening word in order to randomize the data from highly redundant patterns and to reduce DC bias in the packet. Scrambling is performed by a bitwise XOR with the output of a linear feedback shift register (LFSR), defined by the polynomial $D^7 + D^4 + 1$. In the receiver, the inverse processes are carried out. The received data is descrambled using the same whitening word.

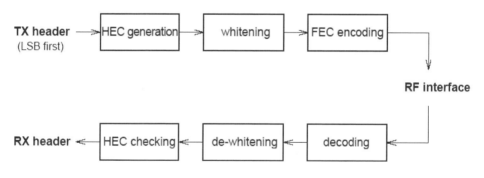

Figure 3–4:   Header bit processing

Figure 3–4 shows the processes carried out for the packet header both at the transmit and the receive side. All header bit processes are mandatory. For the payload, similar processes are performed. The packet type determines which processes are carried out. Figure 3–5 shows the processes that may be carried out on the payload. In addition to the processes defined for the packet header, encryption can be applied on the payload. Only whitening and de-whitening are mandatory for every payload; all other processes are optional and depend on the packet type and the mode enabled. In Figure 3–5, the optional processes are indicated by dashed blocks. After scrambling at the transmitter, FEC encoding is performed. Descrambling is performed at the receiver after FEC decoding.

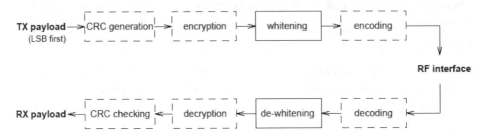

Figure 3–5: Payload bit processing

The piconet is synchronized by the system clock of the master. The master never adjusts its system clock during the existence of the piconet. The slaves adapt their native clocks with a timing offset in order to match the master clock. By comparing the exact RX timing of the received packet with the estimated RX timing, this timing offset is updated each time a packet is received from the master.

Several features are included into Bluetooth to ensure a low-power operation. These features include appropriate signal processing when handling the packets, general architecture, and protocol.

It should be noted that it is much easier for IEEE 802.15.1 to support QoS than it is for IEEE 802.11. The fundamental reason is that IEEE 802.15.1 does not use a CSMA MAC. To emulate full duplex transmission, a Time-Division Duplexing (TDD) scheme is used. The master and slaves transmit alternatively. This controlling mechanism is similar to the point coordination function for WLANs. Recall that the time period for WLANs in which a point coordinator is active is called the contention-free period. In IEEE 802.15.1, because a distributed coordinator is absent, all the time can be considered "contention free". The network is time-multiplexed. The master polls its collection of IEEE 802.15.1 WPAN slaves for transmissions, thus regulating the bandwidth assigned to them. The master can enforce any QoS policy that it desires.

The jitter can be controlled by the packet size. One-slot packets lead to less jitter. In addition, employing a frequency-hopping scheme with the small-sized slots provides noise resilience to interference from other wireless technologies operating in the same band. The master assigns a three-bit active member address to each slave active in its piconet. The all-zero active member address is reserved for broadcast messages. A slave only accepts packets with a matching active member address and broadcast packets.

The master always starts the transmission in the even-numbered slots, whereas the slaves start their transmission in the odd-numbered slots. For greater efficiency, IEEE 802.15.1 uses a combination of circuit and packet switching. Two link types are defined in a piconet—Synchronous Connection-Oriented (SCO) link and Asynchronous Connection-Less (ACL) link.

The connection between two Bluetooth devices is initially an ACL type. The ACL link is a point-to-multipoint link between the master and all the slaves participating on the piconet. In the slots not reserved for the SCO link(s), the master can establish an ACL link on a per-slot basis to any slave, including the slaves already participating in an SCO link. The ACL link provides a packet-switched connection between the master and all active slaves participating in the piconet. If a slave receives an ACL packet from the master in an even-indexed slot, it is allowed to return an ACL packet in the following odd-indexed slot. ACL packets not addressed to a specific slave are considered as broadcast packets and are read by every slave. If the master sends a broadcast message in a even-numbered slot, no slave is allowed to return a packet in the following odd-numbered slot.

Although the ACL link does not have any latency guarantees, a time interval is defined equal to the maximum time between subsequent transmissions from the master to a particular slave on an ACL link. This time interval is called poll interval, T_{poll}, and is used to provide dynamic bandwidth allocation and latency control. To change the bandwidth allocated to a particular ACL connection, the poll interval can be changed. Either the master or the slave can request a new poll interval. This will allow the master and slave to dynamically negotiate the quality of service as needed.

After a connection has been made and the ACL link is established, the master or the slave can request to establish an SCO link by sending a request with parameters that specify the timing, packet type, and coding that will be used on the SCO link. The SCO link reserves slots separated by the SCO interval, T_{SCO}. The first slot reserved for the SCO link is defined by T_{SCO} and the SCO offset, D_{SCO}. After that, the SCO slots follows periodically with the SCO interval. The other device may not accept the suggested SCO parameters and can suggest another possible set of SCO parameters. In this way, the master and the slaves agree on a set of SCO parameters. At any time, the master or the slave can request a change in the SCO parameters or remove the SCO link. The SCO link is a point-to-point symmetric link between a master and a single slave. The master maintains the SCO link by using reserved slots at regular intervals. The SCO link can be considered as a circuit-switched connection because it consists of reserved slots between the master and the slave. Because SCO packets carry time-critical data, they are never retransmitted. The master will send SCO packets at regular intervals to the slave in the reserved master-to-slave slots. The SCO slave will respond with an SCO packet in the following slave-to-master slot. Normally, only the slave that is addressed by the master in a even-numbered slot can return a packet in the next slave-to-master slot. An exception is made for the SCO link to support periodic traffic efficiently. If an SCO link already exists, a slave is still allowed to return an SCO packet in the reserved SCO slot even if the slave is unable to decode its address in the preceding time slot.

For each of the SCO and ACL links, several different packet types are defined. These packet types differ according to the amount of error correction and the number of slots they occupy. There are five packet types common to the ACL and SCO links: ID, NULL, POLL, FHS, and DM1. The ID, NULL, POLL, and FHS packets are link-control packets. The ID packet consists only of the access code and is used in inquiry and paging. The DM1 packet is used for control information on either link and can carry data.

The FHS packet is used to maintain synchronization between the master and all slaves. At connection setup and during a master-slave switch the FHS packet is transferred from the master to the slave. This packet will establish the timing and frequency synchronization. The slaves adjust their RX/TX timing according to the reception of the FHS packet (and not according to the reception of the page message). The FHS packet contains the master's real-time Bluetooth clock, the master's 48-bit Bluetooth device address, and some additional information like the class of device. The FHS packet contains all information to construct the channel access code.

The SCO packets do not include CRC. There are three pure SCO packets. The HV1 packet is protected by a rate 1/3 FEC. The packet length is 240 bits, and because of the FEC, this packet carries 80 information bits. The HV2 packet is protected by a rate 2/3 FEC and has the same overall length of 240 bits. To achieve higher throughput, the HV3 packet is not protected by FEC and carries 240 information bits. If HV1, HV2, or HV3 packets are used for a 64 Kb/s speech signal, one packet will carry 1.25 ms, 2.5 ms, or 3.75 ms of speech. Taking into account the slot period, one packet must be sent every second, fourth, or sixth slot, or T_{SCO} is equal to 2, 4, or 6, correspondingly for HV1, HV2, and HV3. For all of these packets, the hopping rate is 1600 hops/second. Therefore, for a 64 Kb/s speech signal, one HV1, HV2, or HV3 packet is sent correspondingly 800, 400, or 266.6 times each second.

The DV packet is a data-voice SCO packet. This packet carries an asynchronous data field in addition to a synchronous (voice) field. The voice field of the DV packet is 80 bits long and is not protected by FEC. The data field is up to 150 bits long and is protected by a rate 2/3 FEC. The DV packet can support 64 Kb/s for voice, in addition to 57.6 Kb/s for data. In each DV packet, the voice information is new, but the data information might be retransmitted if the previous transmission failed. The parameters of the SCO packets are summarized in Table 3–1.

Table 3–1: SCO packet type parameters

Packet type	HV1	HV2	HV3	DV
Payload header (bytes)	None	None	None	1 data part only
Payload length (bytes)	10	20	30	20
FEC code	1/3	2/3	None	2/3 data part only

There are seven ACL packet types, including the DM1 packet. Six of these packets have 16-bit CRC and can be retransmitted if not acknowledged. These six packet types are DMx and DHx, where x is equal to the slot duration of the packet and can be 1, 3, or 5. DM stands for data-medium rate and DH stands for data-high rate. DM1 has 18 information bytes and a 16-bit CRC protected together by a rate 2/3 FEC. The DH packets are not protected by FEC. The seventh ACL packet type, AUX1, is just like DM1, but has no CRC.

The parameters of the ACL packets are summarized in Table 3–2, and the data rates that these packets achieve are summarized in Table 3–3. Note that for the DM1 and DH1 packets, the data rate for symmetric and nonsymmetric operation is the same. The reason is that these packets are one-slot packets. The use of three-slot and five-slot packets reduces the transmission opportunities (Figure 3–3). The data rates for nonsymmetric operation for DM3, DH3, DM5, and DH5 are obtained by assuming that in one direction three-slot and five-slot packets are used, while in the reverse direction one-slot packets are used. The data rates for symmetric operation are obtained by assuming that multislot packets are used in both directions.

Table 3–2: ACL packet type parameters

Packet type	DH1	DM1	DH3	DM3	DH5	DM5
Slot time	1	1	3	3	5	5
Payload header (bytes)	1	1	2	2	2	2
Payload length (bytes)	0–27	0–17	0–183	0–121	0–339	0–224
FEC code	None	2/3	None	2/3	None	2/3

Table 3–3: Data rate for ACL packets

ACL Packet type	Symmetric	Nonsymmetric
DM1	108.8	108.8/108.8
DH1	172.8	172.8/172.8
DM3	258.1	387.2/54.4
DH3	390.4	575.6/86.4
DM5	286.7	478/36.3
DH5	433.9	721/57.6
AUX1	185.6	185.6/185.6

Note that the data throughput for a given packet type depends on the quality of the RF channel. As an option, quality measurements in the receiver of one device can be used to dynamically control the packet type transmitted from the remote device for optimization of the data throughput.

According to the automatic repeat request scheme, DM, DH and the data field of DV packets are transmitted and retransmitted until acknowledgment of a successful reception is returned or a timeout is exceeded. The cyclic redundancy check (CRC) is used to determine whether the payload is correct or not. The ARQ scheme works only on the payload in the packet and only the payload that has a CRC. The packet header and the SCO payload are not protected by the ARQ scheme. Broadcast packets are checked on errors using the CRC, but no ARQ scheme is applied because the individual slaves never acknowledge broadcast packets. Because broadcast packets are not acknowledged, each broadcast packet is normally repeated for a fixed number of times.

> Retransmissions can be stopped if transmitting the next packet is more important than retransmitting the previous packet. It is clear that the ARQ scheme can cause variable delay in the traffic flow. Also note that although a Bluetooth unit may be in a position where it cannot receive new information, it can still continue to transmit information—the flow control is separate for each direction.

Bluetooth handles voice particularly well. Bluetooth supports three different voice coding formats: µ-law log PCM, A-law log PCM, and CVSD, all at 64 kb/s. The public switched telephone network (PSTN) in the United States, Canada, and Japan uses µ-law log PCM. The PSTN in the rest of world uses A-law log PCM. Because Bluetooth devices do not know which country they are used in, they can perform conversion between A-law and µ-law log PCM. Voice coding can be turned off to achieve a linear PCM data link at 64 kb/s. The 64 kb/s log PCM with either A-law or µ-law companding is done according to ITU-T recommendation G.711. In addition, a sophisticated delta modulator, an innovative type of CVSD, is supported. The modulation scheme follows the waveform where the output bits indicate whether the prediction value is smaller or larger then the input waveform. To reduce slope overload effects, syllabic companding is applied—the step size is adapted according to the average signal slope. The input to the CVSD encoder is linear PCM.

Chapter 3: Standards for Wireless Personal Area Networking (WPAN)

Because there are many voice-compression algorithms, and many of them provide higher compression ratios than CVSD, it is easy to wonder why CVSD was chosen. CVSD was chosen because bit errors are likely, and the quality of the voice depends on the robustness of the voice-coding scheme to bit errors. Compression algorithms achieving high-compression ratios are normally very sensitive to bit errors in the compressed bit stream. In some cases, occasional bit errors can lead to disastrous results. On the other hand, CVSD is rather insensitive to random bit errors. These random bit errors are experienced in voice signals as white background noise. However, when a packet is rejected because either the channel access code or the HEC test was unsuccessful, a whole speech segment will be lost. Additional signal processing can be performed to account for the missing speech parts. This is why the voice payload in the HV2 packet is protected by a 2/3 rate FEC. For errors that are detected but cannot be corrected, the receiver should try to minimize the audible effects at the application layer. For instance, from the 15-bit FEC segment with uncorrected errors, the 10-bit information part as found before the FEC decoder could be used. The HV1 packet is protected by a three-bit repetition FEC. For this code, the decoding scheme will always assume zero or one-bit errors. Thus, no detectable but uncorrectable error events for HV1 packets exist.

Data or link control information is exchanged between the master and the SCO slave using the DV or DM1 packets. Any ACL type of packet can be used to send data or link control information to any other ACL slave.

In order to establish new connections, the procedures inquiry and paging are used. The inquiry procedure enables a unit to discover which units are in range, and what their device addresses and clocks are. The inquiry procedure is used in applications where the other device addresses are unknown. During an inquiry substate, the discovering unit collects the Bluetooth device addresses and clocks of all units that respond to the inquiry message. It can then, if desired, make a connection to any one of them by means of the page procedure. The receiver in the page scan or inquiry scan substate correlates against the inquiry or the device access codes with a matching

> Bluetooth devices have two major states: standby and connection. By default, every Bluetooth unit is in the stand-by state. In this state, the Bluetooth unit is in a low-power mode. The device may leave the standby state to scan for page or inquiry messages, or to perform page or inquiry itself.

correlator. The device access code is used during page, page scan, and page response substates. There is one general inquiry access code (GIAC) for general inquiry operations, and there are 63 dedicated inquiry access codes (DIACs) for dedicated inquiry operations. Various software applications may take advantage of these different inquiry access codes to discover not all devices, but only devices with certain capabilities. For example, one can inquire about printers, facsimile machines, etc. In the page substate, the paging device transmits the device access code (ID packet) corresponding to the slave to be connected rapidly on a large number of different hop frequencies. Because the ID packet is a very short packet, the hop rate is increased from 1600 hops/s to 3200 hops/s. In a single TX slot interval, the paging master transmits on two different hop frequencies. Correspondingly, in a single RX slot interval, the paging receiver listens on two different hop frequencies. This is also necessary, because there could be a large offset between the clocks of the devices. After a successful page attempt, the paging device will enter the connection state as a master, and the paged unit will enter the connection state as a slave.

In the connection state, the connection has been established and packets can be sent back and forth. During the connection state, the Bluetooth units can be in several modes of operation: active mode, sniff mode, hold mode, and park mode.

In the active mode, the Bluetooth unit actively participates on the channel. The hold mode is typically entered when there is no need to send data for a relatively long time. The transceiver can then be turned off in order to save power. But the hold mode can also be used if a device wants to discover or be discovered by other Bluetooth devices or wants to join other piconets. Each device decides what it will actually do during the hold time. The messages exchanged before entering the hold mode contain a parameter, hold instant, which specifies the instant in the future at which the hold will become effective. Before entering the hold mode, the master and slave agree on the time duration the slave will be in the hold mode. The master or the slave can request to enter hold mode. Only the ACL link can be in the hold mode, in which ACL packets are neither transmitted nor accepted. To maintain QoS, SCO links will still be supported. With the hold mode, capacity can be made free to do other things like scanning, paging, inquiring, or attending another piconet. The unit in hold mode may not do any of these things and just be in a low-power sleep mode. During the hold mode, the slave unit keeps its active member address.

The park and sniff modes are similar to the hold mode. When a slave does not need to participate on the piconet channel, but still wants to remain synchronized to the channel, it can enter the park mode, which is a low-power mode with very little activity in the slave. Devices that are in park mode give up their active member address but are assigned a unique parked member address. This parked member address can be used by the master to unpark that slave. To support parked slaves, the master establishes a beacon channel when one or more slaves are parked. The master communicates with parked slaves using this beacon channel. The beacon instants follow periodically at the predetermined beacon interval. Parked slaves sleep most of the time. However, at the beacon instants they wake up to resynchronize to the channel by receiving a packet from the master. Any packet exchanged on the channel can be used for synchronization. The parked slave does not have to wake up at every beacon instant. At the beacon instants, the master can unpark a parked slave, change the park mode parameters, transmit broadcast information to all parked slaves, or let the parked slaves request access to the channel. The master or the slave can request park mode.

> Two parameters control the listening activity in the slave. The sniff attempt parameter determines for how many slots the slave shall listen, beginning at the sniff slot, even if it does not receive a packet with its own active member address. The sniff timeout parameter determines for how many additional slots the slave shall listen if it continues to receive only packets with its own active member address.

Active slaves listen in every master-to-slave slot for packets. If an active slave is not addressed, it may sleep until the next new master transmission period. Sniff mode is used to reduce the duty cycle of the slave. With the sniff mode, the time slots where the master can start transmission to a specific slave are reduced; that is, the master can only start transmission in specified time slots. These so-called sniff slots are spaced regularly with a certain interval, called a sniff interval, T_{sniff}. Before entering a sniff mode, the devices negotiate duration of the sniff interval and a sniff offset, D_{sniff}. The offset determines the time of the first sniff slot; after that the sniff slots follows periodically with the sniff interval. Both slaves and master can request to enter sniff mode. The sniff mode is the least power efficient, followed by the hold mode, and then by the parked mode.

In a scatternet, a device can be simultaneously a master in one piconet and a slave in several other piconets. A master or slave can become a slave in another piconet by being paged by the master of this other piconet. On the other hand, a unit participating in one piconet can page the master or slave of another piconet. Time multiplexing is used to switch between piconets. In case of ACL links only, a unit can request to enter the hold or park mode in the current piconet. During this time, the unit may join another piconet by just changing the channel parameters. Units in the sniff mode may have sufficient time to visit another piconet in between the sniff slots. If SCO links are established, other piconets can only be visited in the nonreserved slots in between. This is only possible if there is a single SCO link using HV3 packets. In the four slots in between, the unit can visit one other piconet.

Because the multiple piconets are not synchronized, guard time shall be left to account for misalignment. This means that only two slots can effectively be used to visit another piconet in between the HV3 packets. Because the clocks of two masters of different piconets are not synchronized, a slave unit participating in two piconets has to take care of two offsets to match time with both master clocks. Because the two master clocks drift independently, regular updates of the offsets are required in order for the slave unit to synchronize to both masters.

Sometimes a master-slave switch is desirable:

- When a unit paging the master of an existing piconet wants to join this piconet because, by definition, the paging unit initially is master of a "small" piconet only involving the pager (master) and the paged (slave) unit.

- When a slave in an existing piconet wants to set up a new piconet, involving itself as master and the current piconet master as slave. This case implies a double role of the original piconet master; it becomes a slave in the new piconet while still maintaining the original piconet as master.

- When a slave wants to fully take over an existing piconet, i.e., the switch also involves transfer of other slaves of the existing piconet to the new piconet, a much more complicated example.

Chapter 3: Standards for Wireless Personal Area Networking (WPAN)

 The master-slave switch requires careful support at the application layer (such as how to handle security and transfer all kind of slave information from old to new master). This procedure is not defined in the 802.15.1 standard, which leaves it to vendors to implement. This is one weakness in the standard, especially from the standpoint of security.

Clearly, this third instance can be achieved by allowing the new master to setup a completely new piconet through the conventional paging scheme. However, that would require individual paging of the old slaves and thus take an unnecessarily long time. Instead, letting the new master use the old master's timing knowledge is more efficient. As a consequence of the master-slave switch, the slaves in the piconet have to be transferred to the new piconet, changing their timing and their hopping scheme.

How does the master-slave switch happen? For the master and slave involved in the role switch, it reverses their TX and RX timing—a TDD switch. Moreover, because the piconet parameters are derived from the device address and clock of the master, a master-slave switch inherently involves a redefinition of the piconet as well—a piconet switch. Assume unit A wants to become master; unit B was the former master. Then there are basically two alternative master-slave switch scenarios—either the slave takes the initiative or the master takes the initiative.

First, both slave A and master B do the TDD switch but keep the former hopping scheme (still using the device address and clock of unit B); there is no piconet switch yet. Unit A now becomes the master, unit B, the slave. The FHS packet is sent by master A using the "old" piconet parameters. The active member address in the FHS packet header is the previous address used by unit A. From this moment, Slave B starts to use the active member address formerly used by unit A in its slave role. After the FHS acknowledgment, which consists of the ID packet and is sent by the slave on the old hopping sequence, both master A and slave B turn to the new channel parameters of the new piconet as indicated by the FHS packet.

Bluetooth security

WPANs are ad hoc, and security is an important requirement. To provide security, measures must be taken both at the application layer and the data-link layer.

The goal of the security mechanism at the data-link layer is to provide efficient support for different applications. To be appropriate for ad-hoc networks—i.e., for a peer environment—the authentication and encryption routines in every unit must be implemented in an identical fashion. Four different entities are used for maintaining security at the data-link layer: a public address that is unique for each user; two secret keys; a a random number, which is different for each new transaction.

The public address is the 48-bit Bluetooth device address, which is unique for each Bluetooth unit. The Bluetooth addresses are publicly known and can be obtained, for example, from an inquiry by a Bluetooth unit.

The two secret keys are derived during initialization and are never further disclosed. One is used for authentication and called authentication key, the other is used for encryption and called encryption key.

The fourth entity, the random number, can be derived from a random or pseudorandom process in every Bluetooth unit. It is not a static parameter and changes frequently. Random numbers are used for many purposes within the security functions, e.g., for the challenge-response scheme, for generating authentication, and encryption keys. Within Bluetooth, the requirements placed on the random numbers used are that they be nonrepeating within the lifetime of the authentication key and randomly generated.

First consider first the authentication key and then the encryption key. The length of the authentication key is always 128 bits. The authentication key is also referred to as the link key. This link key is shared between two or more parties and is the base for all security transactions between these parties. The current link key is used for all authentications and all generation of encryption keys in the on-going session. Because piconets are ad-hoc networks, every unit participates in a piconet for a finite time interval. This time interval is called a session. Thus, the session terminates when the unit disconnects from the piconet. The link keys are either semipermanent or temporary. A semipermanent link key is stored in nonvolatile memory and may be reused after the current session is terminated. The lifetime of a temporary link key is limited by the lifetime of the current session; it cannot be reused in a later session. Actually, there are four types of authentication (or link) keys:

- The unit key K_A
- The combination key K_{AB}
- The temporary key K_{master}
- The initialization key K_{init}

For a Bluetooth unit, the combination key K_{AB} and the unit key K_A are functionally indistinguishable; the difference is in the way they are generated. The unit key, K_A is generated in a single unit, which can be denoted as unit A. The combination key K_{AB} is derived from information in, for example, two units A and B and is therefore always dependent on two units. The combination key is derived for each new combination of two Bluetooth units. Whether a unit key or a combination key is used depends on the application. If a Bluetooth unit has little memory to store keys or has to communicate with a large group of users, it can use its own unit key. In that case, it has to store only a single key. Preferably, applications that require a higher security level will use the combination keys. These applications will require more memory because a combination key for each link to a different Bluetooth unit has to be stored.

The initialization key, K_{init}, is used as link key during the initialization process when no link key has been defined or when a link key has been lost.

Typically, in a point-to-multipoint configuration where the same information is to be distributed securely to several recipients, a common key is useful. To achieve this, a special link key (denoted master key) can temporarily replace the current link keys. The master key, K_{master}, is a link key only used during the current session.

The link (or alternatively, authentication) keys must be generated and distributed among the Bluetooth units in order to be used in the authentication procedure. Because the link keys must be secret, they cannot be obtained through an inquiry routine in the same way as the Bluetooth addresses. The exchange of the keys takes place during an initialization phase. All initialization procedures consist of the following five parts: (1) generation of an initialization key; (2) generation of link key; (3) link key exchange; (4) authentication; (5) generating of encryption key in each unit (optional).

The first step is the generation of an initialization key. This initialization key can be derived, for example, from a random number, the Bluetooth device address, and user-specified PIN code. The initialization key is used temporarily. After the units have performed the link key exchange, the initialization key is discarded. A fraudulent Bluetooth unit may try to test a large number of PINs by claiming another Bluetooth device address each time. It is the application's responsibility to take countermeasures against this threat. This initialization is carried out separately for each two units that want to implement authentication and encryption.

The second and third steps are the generation and exchange of link keys. At initialization, the two applications running at the two devices must determine which key will be used as the link key—K_A, K_B, or K_{AB}. (Which key will be selected is not specified in 802.15.1.) Typically, this will be the unit key of the device with restricted memory capabilities, because this unit only has to remember its own unit key. If this unit key is chosen, then this unit key is transferred to the other party and stored as link key for that particular party. If it is desired to use the combination key, this key is first generated during the initialization procedure. The combination key is the combination of the two random numbers generated in units A and B. First, each unit generates a random number and then the two units exchange the two random numbers that they have generated.

> The unit keys K_A and K_B are generated when the Bluetooth unit is in operation for the first time; i.e., not during each initialization. Once created, these unit keys are stored in nonvolatile memory and (almost) never changed. For example, if one of these unit keys is being used as the link key for a particular pair of devices, and this key is changed after initialization, the other device will possess a wrong link key.

When the random numbers and have been mutually exchanged, each unit recalculates the other unit's contribution to the combination key. This is possible because each unit knows the Bluetooth device address of the other unit. The result is stored in unit A as the link key for the connection with device B, and in unit B as the link key for the connection with device A. For any new connection established between units A and B, they will use the agreed upon link key, instead of once again going through this procedure. However, if no link key is available, the devices will perform the initialization procedure again.

After the link key exchange, the fourth step—mutual authentication—can be initiated. The authentication used in Bluetooth uses a challenge-response scheme. In this algorithm, a claimant's knowledge of a secret key is checked through a two-step protocol using symmetric secret keys. Authentication succeeds if the claimant and the verifier share the same secret key. In the challenge-response scheme, the verifier challenges the claimant to authenticate a random input (the challenge), with an authentication code and return the result to the verifier. The claimant calculates a response (which is a function of the challenge), the claimant's Bluetooth device address, and a secret key. The response is sent back to the verifier, which checks if the response was correct. Successful calculation of the authentication procedure requires that two devices share a secret key. Both the master and the slave can be verifiers. Mutual authentication is achieved by first performing the authentication procedure in one direction and then immediately by performing the authentication procedure in the opposite direction. Certain applications only require a one-way authentication. However, in some peer-to-peer communications, one might prefer a mutual authentication in which each unit is subsequently the challenger (verifier) in two authentication procedures. In general, the application indicates who has to be authenticated by whom. As a side effect of a successful authentication procedure, an auxiliary parameter (the Authenticated Ciphering Offset [ACO]) will be computed. This ACO is used in the generation of the encryption key. If an authentication is successful, the value of ACO should be retained. When the authentication attempt fails, a certain waiting interval must pass before the verifier can initiate a new authentication attempt to the same claimant or before it will respond to an authentication attempt initiated by a unit claiming the same identity as the suspicious unit. For each subsequent authentication failure with the same Bluetooth address, the waiting interval is increased exponentially. The waiting interval is limited to a certain maximum. The maximum waiting interval depends on the implementation. The waiting time shall exponentially decrease to a minimum when no new failed attempts are being made during a certain time period. This procedure prevents an intruder from repeating the authentication procedure with a large number of different keys. To make the system somewhat less vulnerable to denial-of-service attacks, the Bluetooth units should keep a list of individual waiting intervals for each unit with which it has established contact.

The fifth step is optional and is performed only if encryption is turned on.

After the initialization procedure, the units can proceed to communicate, or the link can be disconnected. If encryption is implemented, the E_0 algorithm is used with the proper encryption key derived from the current link key.

The length of the encryption key can vary between 8 and 128 bits. The size of the encryption key is configurable for two important reasons—one political and one technical. The first reason is the government regulations regarding the export of cryptographic algorithms. The technical reason to make the encryption key of variable length is to provide an easy upgrade path for future hardware and software implementations.

Encryption can be turned on and off many times, and each time encryption is activated, a new encryption key is derived from the current link key. Thus the lifetime of the encryption key does not necessarily correspond to the lifetime of the authentication key. The authentication key is more static than the encryption key. Only the particular application running on the Bluetooth device decides when, or if, to change it. The purpose of separating the authentication key and encryption key is to facilitate the use of a shorter encryption key without weakening the strength of the authentication procedure. Note that there are no governmental restrictions on the strength of authentication algorithms.

> According to U.S. government regulations, the export of technology with strong encryption outside the United States and Canada is not allowed. Increasing the effective key size is the simplest way to combat the increased computing power of attackers. Furthermore, government regulations do not allow users to be able to set the encryption key size. Thus, although the encryption key length is variable, it should be a factory preset entity. Therefore, the Bluetooth baseband processing does not accept an encryption key given from higher software layers.

The encryption key, K_C, is derived from the current authentication key, a 96-bit Ciphering Offset number (COF), and a 128-bit random number. This COF is either derived from the Bluetooth device address of the master or is set to the value of ACO as computed during the authentication procedure. The encryption key is automatically changed each time the unit enters the encryption mode.

It is quite possible for the master to use separate encryption keys for each slave in a point-to-multipoint configuration with encryption activated. However, if the application requires more than one slave to listen to the same payload, each slave must be addressed individually. This clearly reduces throughput. Moreover, a Bluetooth unit cannot be expected to be capable of switching between two or more encryption keys in real time. Thus, the master cannot use different encryption keys for broadcast messages and individually addressed traffic. Instead of addressing each slave individually, even for broadcast messages, the master may tell several slave units to use a common link key. This key is called K_{master}. Because the encryption key is derived from the link key, the encryption key will be common as well. However, this common link key is only of temporary interest because many applications use broadcast messages only part of the time.

To create a broadcast group, the master will issue a command to the slaves to replace their respective current link key by the new (temporary) master key. Before encryption can be activated, the master also has to generate and distribute a common random number to all participating slaves. Using this random number and the newly derived master key, each slave can generate independently the new common encryption key. Then the master can broadcast encrypted information. Note that because the encryption key is common, all the slaves in the broadcast group must support the length of this encryption key. Those slaves that cannot must be excluded from the broadcast group. The broadcast group must be a trusted domain, because each slave in possession of the master key can eavesdrop on all encrypted traffic, not only on the traffic intended for itself. If there is no more broadcast traffic or if there is a change in the security policy, the master can tell all participants to fall back to their old link keys simultaneously.

In certain circumstances, modifying the link keys is desirable. A link key based on a unit key can be changed, but not very easily. The unit key is created once during the first use. Changing the unit key is a less desirable alternative, because several units may share the same unit key as link key. One example is a printer whose unit key is distributed among all users using the printer's unit key as a link key. Changing the unit key will require reinitialization of all units trying to connect. In certain cases, this might be desirable; for example, to deny access to previously allowed units. Another possibility is starting up an entirely new initialization procedure. In that case, user interaction may be necessary because a PIN is required in the authentication and encryption procedures.

The encryption procedure is discussed next. User information can be protected by encryption of the packet payload; the access code and the packet header are never encrypted. The encryption of the payloads is carried out with a stream cipher called E_0 that is resynchronized for every payload. The overall principle is shown in Figure 3–6.

The stream cipher shown in Figure 3–6, consists of three steps. The first step is the generation of the payload key. The payload key is constructed from the master Bluetooth address, 26 bits of the master real-time clock and the encryption key as input. The encryption key, K_C, is derived from the current link key, COF, and a random number. The master transmits the random number before entering encryption mode. This random number is publicly known because it is transmitted as plain text over the air. The algorithm is reinitialized at the start of every new packet. By using the real-time clock, at least one bit is changed for every new packet. The payload key generator is very simple—it merely combines the input bits in an appropriate order.

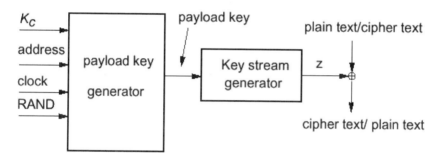

Figure 3–6: A block diagram of the E_0 stream cipher

The main part of the cipher algorithm is the second step. The second step generates the key stream bits, and the third step performs the encryption or decryption. The actual encryption is performed by adding the ciphering bits bit-wise modulo-2 to the data bits. The payload is ciphered after the CRC bits are appended, but before the FEC encoding. Each packet payload is ciphered separately. The cipher is symmetrical; decryption is performed in exactly the same way using the same key as used for encryption.

Because the encryption key is of variable length, each Bluetooth device has a parameter defining the maximal allowed key length. Every application must define a number indicating the smallest acceptable key size for that particular application. Before generating the encryption key, the involved units must negotiate to decide what key size to actually use. This negotiation can happen by the master sending a suggested value and the slave deciding whether to accept it or not. If the suggested key length is not acceptable to the slave, it will in turn suggest a different key length. This procedure is repeated until a key length agreement is reached, or, under the control of the application one unit aborts the negotiation. If the agreed key length is too short for the application, the units cannot exchange encrypted information.

The master and every slave in the piconet must agree first whether to use encryption or not. If encryption is used, they must agree whether to use encryption only on the point-to-point packets or on both the unicast and multicast packets. If a slave has a semi-permanent link key (i.e., a combination key or a unit key), it can receive encrypted data only on slots individually addressed to it. In particular, it will assume that broadcast messages are not encrypted.

LMP and L2CAP

The protocol used for link set-up, security, and control is called link manager protocol (LMP). The place of the LMP in the protocol stack is illustrated in Figure 3–7. LMP sits on top of the baseband protocol. LMP messages are transferred in the payload and are distinguished by a reserved value in the payload header. The LMP messages are exchanged between LMP entities and are not propagated to higher layers. Link manager messages have higher priority than user data. LMP messages are not acknowledged. LMP messages are used when devices enter hold, park, or sniff modes. The LMP also performs power control. If the RSSI value differs too much from the preferred value of a Bluetooth device, it can request an increase or a decrease of the other device's TX power. If a device does not support power control requests, this is indicated in the supported features list. Upon receipt of this message, the output power is increased or decreased one step, where the step size is defined in the radio specification. At the master side, the TX power is completely independent for different slaves; a request from one slave can only affect the master's TX power for that same slave.

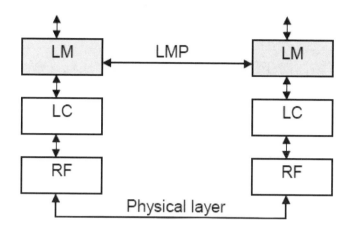

Figure 3-7: The LMP in Bluetooth protocol stack

The link controller (LC) cannot guarantee that it will deliver a message to the remote device within a certain period of time. Therefore, there can be no guarantees that LMP messages will be delivered within a certain period of time.

Each Bluetooth link has a timer that is used for link supervision. This timer is used to detect link loss caused by devices moving out of range, a device's power-down, or other similar failure cases.

The Link Manager plays a role in setting up ACL and SCO links, etc. The LMP plays a role in the paging mechanism. In addition to the mandatory paging scheme, Bluetooth defines optional paging schemes. LMP provides a means to negotiate the paging scheme, which is to be used the next time a unit is paged.

Another protocol of the Bluetooth specification is the logical link control and adaptation protocol (L2CAP). This protocol supports higher-level protocol multiplexing, packet segmentation and reassembly, and the conveying of quality of service information. Figure 3-8 illustrates how L2CAP fits into the Bluetooth Protocol Stack. L2CAP lies above the Baseband Protocol and interfaces with other communication protocols such as the Bluetooth Service Discovery Protocol (SDP), RFCOMM, and Telephony Control (TCS). Voice-quality channels for audio and telephony applications are usually run over Baseband SCO links. Also, packetized audio data, such as IP Telephony, may be sent using communication protocols running over L2CAP.

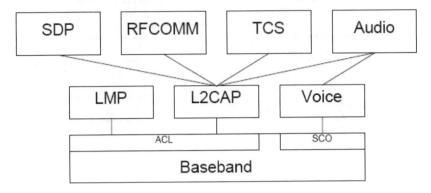

Figure 3-8:   L2CAP in the Bluetooth protocol stack

COEXISTENCE AMONG WIRELESS STANDARDS
Overview

The major advantage of wireless communication devices is that they do not require wires. One disadvantage is that, well, because they do not require wires different wireless devices can operate in the same frequency band. Multiple-access methods have been thoroughly investigated, but until recently no one studied the operation of several different wireless systems in the same frequency band. Multiple-access methods consider how multiple devices can use the spectrum according to the same protocol. They do not (and cannot) consider how different wireless systems can coexist in the same frequency band. Every wireless system is designed assuming that it has the entire spectrum in which it operates available to itself. What will happen if there are two or more different wireless systems in the same frequency and deployed in proximity to each other?

Devices that transmit at a relatively higher power and/or use protocols more robust to interference will get their data through. The other devices will not get their data through, or at least their throughput will be reduced. This situation can be called the "big stick" policy. Devices that have a bigger stick will work. It is obvious that as the number of wireless devices based on different standards grows,

Chapter 3: Standards for Wireless Personal Area Networking (WPAN)

the lack of coexistence among the devices can discourage consumers to use more wireless devices.

This issue seems important in home networking applications, where the number of wireless devices is expected to skyrocket in the next several years. Therefore, measures need to be taken to ensure that different wireless systems will be able to function, even when they use the same frequency band. In other words, replace the "big stick" policy with the "good citizen" policy. The IEEE 802—more specifically, the IEEE 802.15 Working Group—is the first standards-making body to consider this problem.

Before the recommended practice for coexistence is discussed, the meanings of coexistence, interoperability, interworking, and backwards compatibility must be defined. *Coexistence* is the ability of one system to perform a task in a given shared environment in which there are other systems that may or may not be using the same set of rules. *Interoperability* is the ability of two systems to perform a given task using a single set of rules. The ability of two systems to perform a task when each system implements a different set of rules is called interworking. If a new system can interwork with an old system, the new system is backwards compatible with the old system.

Many parameters affect the interference between two different wireless systems, such as separation among antennas, amount of data traffic, power levels, and modulation types. These are physical layer parameters. However, it is important to note that interference is not only a physical layer problem. In most of the previous research, the role of the MAC layer in interference and coexistence problems was not considered. Not only the MAC protocol can play a very important part in the interference scenario and affect the overall system performance, but the reverse is also true—interference can significantly impact the performance of the MAC layer and higher layers.

The IEEE 802.15.2 task group was formed to develop a recommended practice for coexistence between Bluetooth/IEEE 802.15.1 and IEEE 802.11. When IEEE 802.15.2 was formed, it was expected that WPAN devices based on Bluetooth/IEEE 802.15.1 and WLAN devices based on IEEE 802.11b will have the largest market share of devices operating in the 2.4 GHz ISM band. The recommended practice considers the IEEE 802.11b direct-sequence spread-spectrum standard at data rates of 1, 2, 5.5, and 11 Mb/s. Both IEEE 802.11 and

IEEE 802.15 are continuing to work on additional standards. Future revisions of the recommended practice may consider the coexistence issue among those newer WLAN and WPAN standards.

For IEEE 802.11, the data rate typically will be adjusted by the application to be the maximum data rate that results in a packet error rate of 10 percent. If the number of bits per packet is N, the packet error rate can be determined from the bit error rate (BER) as

$$PER = 1 - (1 - BER)^N \qquad \text{Eq. 3–1}$$

At larger distances as the packet error rate increases the data rate will decrease. IEEE 802.11b is occupying a 22 MHz-wide band within the 2.4 GHz ISM band. IEEE 802.15.1 hops on 79 frequencies within the entire band. There would be a collision if IEEE 802.15.1 hops on a frequency occupied by IEEE 802.11b. Out of the 79 frequencies, 22 are subject to interference. The decrease of the data rate by the application can lead to instability. For example, due to interference from IEEE 802.15.1, it is possible for the application to lower the data rate of IEEE 802.11. At a lower data rate, however, packets become longer and can experience interference for longer periods of time, leading to increase in interference, and so on.

In its turn, IEEE 802.11b can cause interference to IEEE 802.15.1, affecting 22 out of the 79 channels and typically causing throughput decrease.

Note that transmitter power control and rate scaling is a simple coexistence mechanism. Power control can be effective if IEEE 802.11b and Bluetooth/IEEE 802.15.1 devices are designed to limit their transmit powers near the threshold to obtain the required performance. To implement power control, a transmit power control mechanism must be implemented. All IEEE 802.11b devices currently support multiple transmit rates, i.e., 1, 2, 5.5 and 11Mb/s. As a result, all IEEE 802.11b devices currently implement a rate shifting/control algorithm. Systems monitor SNR, SINR, PER, etc., to determine which rate should be used. The maximum rate is always desired. The rate is shifted down when packets cannot be successfully decoded at current rate. The rate control algorithm of IEEE 802.11b devices can be extended to incorporate the highest mandatory rate at lower transmit powers. The rate shift algorithm would shift to the highest possible rate with lower transmit power whenever possible.

The recommended practice, developed by the IEEE 802.15.2 Task Group [B79], has developed more sophisticated coexistence approaches. It starts with a detailed computer model of the mutual interference between IEEE 802.11 WLAN and IEEE 802.15.1 WPAN. The parameters of this model are the PHY and MAC models, network topology, user density, propagation model, and data traffic model. The coexistence model quantifies the effect of mutual interference of WLAN and WPAN on each other. This computer model is based on detailed simulation models of the RF channel, and the PHY and MAC of the standards. The recommended practice is considering only the DCF channel access mechanism for IEEE 802.11.

The remainder of this subsection will provide an overview of the coexistence mechanisms, followed by a discussion on the collaborative coexistence mechanisms, which, in turn is followed by a discussion on noncollaborative approaches. Because channel classification is one important part of the coexistence mechanism, it is considered in "Channel classification" on page 174. The analytic and computer MAC and PHY models are discussed in "PHY and MAC models" on page 176.

After deliberation, the IEEE 802.15.2 Task Group developed two types of coexistence mechanisms—collaborative and noncollaborative. In collaborative mechanisms, WPAN and WLAN exchange information over a separate link between one another to minimize mutual interference. The type of link and the protocol for communication over the link are not specified in the recommended practice. This is an opportunity for additional proprietary firmware and software for products, which support both IEEE 802.11b and Bluetooth. Collaborative coexistence mechanisms can be implemented when, for example, one device is a member of both the IEEE 802.11b and IEEE 802.15.1 networks. Noncollaborative coexistence mechanisms do not require exchange of information between the two networks. A total of eight coexistence mechanisms are specified in the recommendation—three collaborative and five noncollaborative.

The first collaborative method is simple. If the two networks have a separate link, then their transmissions can be scheduled in different time segments. This technique is called alternating wireless medium access.

The second collaborative method is packet traffic arbitration. In this method, the IEEE 802.11b and IEEE 802.15.1 networks can decide what type of packet to send to minimize the interference between them. The first and the second collaborative mechanisms require joint action by both networks.

The third collaborative mechanism is to perform deterministic interference suppression by IEEE 802.11b. In this method, IEEE 802.11b uses a programmable notch filter to filter out the narrowband IEEE 802.15.1 interference. Clearly, no action is required by the IEEE 802.15.1 devices.

The first noncollaborative method is the IEEE 802.11b device to perform adaptive interference suppression. In this case, the notch filter must be adaptive to filter out the narrowband IEEE 802.15.1 interference. In this method no action is required on the part of the IEEE 802.15.1 devices.

There are four noncollaborative methods, in which no action on the side of IEEE 802.11b is required. In these four methods, only the IEEE 802.15.1 network takes measures to avoid or minimize interference. These methods are adaptive packet selection, packet scheduling on the ACL link, packet selection on the SCO link, and adaptive frequency hopping.

Collaborative mechanisms are recommended, when possible. If many devices are close together and the potential for interference is high, the alternate wireless medium access mechanism is recommended. If the potential for interference is low, the packet scheduling, combined with deterministic interference suppression is recommended.

Finally, it must be noted that two or more coexistence mechanisms can be used simultaneously. Again, this is not specified in the recommended practice and is an area for developing intellectual property. Some vendors regard the coexistence performance as an area of product differentiation in the marketplace.

Collaborative methods

The first collaborative method is alternating wireless medium access. Recall that an IEEE 802.11b access point sends out a beacon at approximately periodic time intervals. According to this coexistence method, a portion of the IEEE 802.11b beacon period is used for WLAN traffic, and a portion is used for IEEE 802.15.1 traffic. Information about the length of the WLAN and WPAN periods is included in the beacon. Because the two networks transmit in nonoverlapping time periods, there is no interference between them. This method clearly requires synchronization. To achieve synchronization between the networks, an IEEE 802.11 station must send a synchronization signal to the IEEE 802.15.1 master. This is simple to accomplish if the master of the IEEE 802.15.1 piconet and an IEEE 802.11b station are in the same device. If ACL packets are used, it can be guaranteed that WPAN devices do not transmit during the WLAN portion of the beacon period, because a slave is not allowed to send an ACL packet if it is not addressed the previous time slot. However, a slave is allowed to send an SCO packet, even if it is not addressed the previous slot. As a result, there can be no guarantee that an SCO packet will not be sent during the time allocated for WLAN. Therefore, this coexistence method is not suitable for situations with SCO connections.

The second collaborative method is packet traffic arbitration (PTA). In this method, the IEEE 802.11b station and the IEEE 802.15.1 unit are collocated. Every packet about to be transmitted is submitted to a block called the packet traffic arbitrator for approval or rejection. If the transmission will not result in collision between the networks, the packet traffic arbitrator grants approval. If the requests from the two networks are simultaneous or almost simultaneous and there would be a collision, prioritization is used. One of the requests will be approved, and the other will be delayed. Priority could be assigned in a deterministic or random fashion. The deterministic approach is to assign priority levels in the following order from highest to lowest: IEEE 802.11b acknowledgment packets, IEEE 802.15.1 SCO packets, IEEE 802.11b data packets, and IEEE 802.15.1 ACL packets. Furthermore particular implementations can use different fairness criteria such as PER or BER to grant or approve requests. Both SCO and ACL traffic are supported, and SCO traffic is given higher priority than the ACL traffic in scheduling. The PTA method is somewhat similar to alternating wireless medium access, in the sense that only one of the networks transmits at any given time. However, medium access is divided dynamically.

Deterministic interference suppression is the third collaborative mechanism. It is based on the following idea. Because the bandwidth of IEEE 802.15.1 is only 1 MHz, it can be considered as a narrowband interferer to IEEE 802.11b, which has bandwidth of 22 MHz. A notch filter can suppress the IEEE 802.15.1 signal. This notch filter is placed at the IEEE 802.11b receiver. Because the IEEE 802.15.1 signal is hopping within the entire band, the notch filter must be programmable and must hop synchronously with IEEE 802.15.1. Therefore, the IEEE 802.11b receiver must have an integrated IEEE 802.15.1 unit to provide knowledge of the hopping sequence and timing. Deterministic interference suppression works only at the physical layer, and prevents interference from IEEE 802.15.1 to IEEE 802.11b, but not vice versa.

Noncollaborative methods

If an IEEE 802.11b device is not colocated with an IEEE 802.15.1 device, it is not possible to use collaborative coexistence methods. While deterministic interference suppression is not possible, adaptive interference suppression is possible. It is again entirely based on signal processing to suppress IEEE 802.15.1 interference to IEEE 802.11b. Because the IEEE 802.11b receiver does not know the hopping sequence and the associated timing, this method requires an adaptive filter.

The adaptive packet selection and scheduling, another noncollaborative mechanism, is possible because IEEE 802.15.1 systems use various packet types with different packet lengths and degree of error protection. The interference between IEEE 802.15.1 and IEEE 802.11b depends on the type of packet used. For instance, if the channel is dominated by interference from IEEE 802.11b network, packet loss will be mainly due to collisions between IEEE 802.15.1 and IEEE 802.11 systems, instead of bit errors resulting from noise. Note that FEC methods by design provide better protection against errors resulting from noise sources. However, FEC is not effective against interference. Therefore, packet types that do not include FEC will be shorter and could provide better throughput.

The key idea is to adapt the transmission according to channel conditions. Note the difference with PTA method. Because packet traffic arbitration is a collaborative mechanism, it does not have to perform channel estimation. Adaptive packet selection is noncollaborative and relies on channel estimation. Channel estimation can be done in a variety of ways. For example the RSSI, BER, and PER all provide information about the channel. It is also possible to combine these measures. The recommended practice does not specify how channel estimation can be done.

This coexistence mechanism can be considered as an optimization problem. Given the channel conditions, select first the packet type, and then the time of packet transmission so that interference is minimized and the resulting throughput is within a certain range. One simple solution to this optimization problem is to adjust the time of packet transmission so that the IEEE 802.15.1 devices transmit during hops outside the WLAN frequencies and refrain from transmitting while in-band. This method does not require any action on the part of the IEEE 802.11b devices.

The first algorithmic problem is adaptive packet selection. Bluetooth specifies a variety of packet types with different combinations of payload length, slots occupied, FEC codes, and ARQ options. Choosing the packet type that will achieve maximum network performance can be tricky. First, consider an SCO link. Bluetooth provides four types of packets that can be sent over an SCO link: HV1, HV2, HV3, and DV packet. The different packets differ mostly in the FEC code that is used. The ACL link supports six packet types. They differ in the coding rate and the number of occupied slots. Again, the different SCO and ACL packets allow trade-offs to be made mainly between throughput and generated interference. Within this trade-off will be the solution to the optimization problem.

Consider first the case of a range-limited SCO connection. A receiver can decide that a connection is range-limited by low values of the RSSI. In a range-limited connection, random bit errors, not interference, are the dominant problem. In this case, choosing a packet type with stronger error-correcting capabilities will increase the performance of the SCO link. Therefore, for range-limited applications, HV1 packet is preferred over HV2 packet, and HV2 packet is preferred over HV3 packet.

In the case of interference-limited SCO connection, the dominant source of bit errors is the interference produced by the IEEE 802.11b network. A receiver can decide that a connection is interference-limited by high values of the RSSI and high BER, for example. In this case, lowering the coding rate to increase the protection will cause the Bluetooth device to generate more packets, because of the additional parity bits that must be sent. In turn, this creates more interference to the IEEE 802.11b network and ultimately instability, significantly decreasing the total network throughput. This unstable situation will end when the hopping frequency is out of the IEEE 802.11b band. Therefore, in interference-limited scenarios (as in Bluetooth and IEEE 802.11b coexistence scenarios), HV3 packets will lead to better results over HV2 packets, and HV2 packets over HV1 packets.

For the ACL packets, the situation is similar. When the Bluetooth network performance is range-limited, ACL packets with FEC protections (which include DM1, DM3, and DM5) will obtain better results. On the other hand, when the system is interference-limited, bandwidth efficient packets such as DH1, DH3, or DH5 will produce better results.

The third noncollaborative technique is ACL packet scheduling. This approach consists of two components: (1) frequency classification and (2) master delay policy. In the frequency classification phase, the Bluetooth devices measure channel conditions on the different frequencies and classify the frequencies as "good" or "bad." Again, the actual method for channel classification is not specified, but RSSI, BER, and PER are good criteria. The results are entered into a frequency classification table for every receiver. These tables should be kept at the master of the piconet. The goal of the so-called master delay policy is to avoid a packet transmission in a "bad" receiving channel. Because the master controls the network this policy must be implemented only at the master. The basic idea is to wait for a "good" frequency in the frequency-hopping pattern. Furthermore, because following each master's transmission, there is a slave transmission, the master checks both the slave's receiving frequency and its own receiving frequency before choosing to transmit a packet in a given frequency hop.

This scheduling policy is effective in reducing packet loss and delay, especially for multislot Bluetooth packets. In addition, packets are not transmitted when the channel is bad, which saves power. Also, channels occupied by other devices are avoided, which eliminates interference to other systems. Therefore, scheduling is a "good-citizen" policy. However, it must be noted that scheduling works only for ACL packets, because SCO packets must be sent at fixed intervals.

> To further increase the robustness of the ACL link, the adaptive frequency hopping mechanism can be used in conjunction with a packet-scheduling algorithm.

Another noncollaborative mechanism is adaptive frequency hopping. It is a method according to which the hopping sequence is dynamically modified according to the frequency classification table, and devices hop over only the good frequencies. Therefore, channel classification and channel classification exchange between the master and all the active slaves on the piconet is a prerequisite for adaptive frequency hopping.

Adaptive frequency hopping (AFH) (Figure 3–9) consists of the three distinct components:

- The legacy hop kernel
- The partition sequence generator
- The frequency remapping function

The first component of the AFH mechanism is the legacy hop kernel, which generates the hopping sequence as defined in IEEE 802.15.1. The second component of the AFH mechanism is the partition sequence generator. Partition mapping is a method to select one channel from the partition. Each channel in the partition should have the same probability of selection on average. The mapping is pseudorandom. The pseudorandom mapping device maps the selected channel from the original hopping sequence into one channel of the partition specified by the partition sequence. The partition sequence generator makes sure that the new hopping sequence simultaneously appears to be random and is within the set of the good channels.

The output of the partition sequence generator is then used as an input to the final component of the AFH mechanism, the frequency remapping function, which generates an adapted hopping sequence with the appropriate structure. The frequency remapping function remaps (if necessary) the hopping frequency produced by the legacy kernel on to the set of the good channels S_G or the set of bad channels to be kept S_{BK}. Note that when the input to the frequency remapping function is a constant signal of one, it produces an adapted hopping sequence that only hops over the good channels. The remapping of the hopping frequencies is fairly straightforward. If the legacy hopping frequency is already in the set that is specified by the partition sequence, then the output of the frequency remapping function is equal to the legacy hopping frequency. However, if the legacy hopping frequency is not in the required set, then the legacy hopping frequency is remapped to a different frequency.

AFH would work well as long as the number of the good channels is above a certain minimum. Note that in the United States and other countries, regulatory bodies have set a limit on the minimal number of channels in the hopping sequence. System robustness also requires a minimal number of hopping channels. Depending on the number of available good channels, some bad channels may have to be placed in the hopping sequence. If the number of good channels is below the minimum, first the SCO packets are allocated to the good channels, and what is left is distributed among the ACL packets. A level of QoS is thus guaranteed in interference environment.

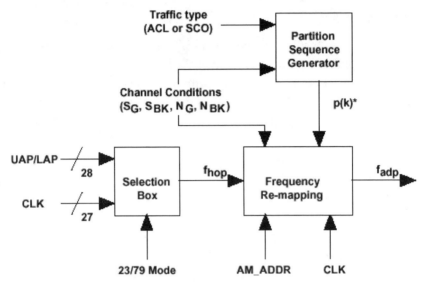

Figure 3-9: Block diagram of the adaptive frequency-hopping mechanism

Channel classification

Channel classification is required in both of the two noncollaborative mechanisms. Adaptive packet selection and scheduling adapts the packet types and transmission timing to the channel condition of the current hopping channel. Adaptive frequency hopping generates the new hopping sequence based on the result of channel classification. The goal of channel classification is to determine the quality of each channel. As discussed earlier, channels experiencing a high degree of interference are classified as "bad" channels, and channels with low interference are classified as "good" channels. The channel classification mechanism is not specified in the standard and is left to the technology and equipment vendors to implement. There are three sources of errors, each one leading to a packet error: access code error, header error, and payload error. In the case of access code error, the packet will not be detected. Because it does not know what was transmitted, an actual receiver cannot measure exactly the packet error rate. However, a receiver can compute the ratio between the number of packets with a CRC error (error in the payload) to the number of correctly received packets.

High values for the RSSI and the PER indicate that the errors are due to interference. A low value of the RSSI might indicate that the distance between the devices is significant, but it would not be a reason for classifying a channel as "bad." Every slave and the master perform channel classification. Suppose that S_{ij} is the assessment by slave i of channel j, $i = 1,\ldots N_s$, where N_s is the number of slaves, and $j = 1,\ldots N_c$, N_c being the number of channels (79 or 23, depending on the regulatory domain). The master's assessment of channel j is M_j. S_{ij} and M_j take only binary values, 1, if channel j is assumed to be good, and 0, if it is assumed to be bad. The master uses the classification lists from the slaves and its own classification list to produce a final classification list. The quality of the j-th channel is assumed to be:

$$Q_j = \frac{M + \sum_{i=1}^{N_s} S_{i,j}}{N_s + 1} \qquad \text{Eq. 3–2}$$

$$1 \leq N_s \leq 7$$
$$0 \leq j \leq N_c$$

where the quality of one channel is a linear combination of the assessments of the master and the slave,

$$W_{ij} = \alpha M_j + (1 - \alpha) S_{ij} \qquad \text{Eq. 3–3}$$

To ultimately classify channel j as good or bad, a threshold is applied to the value of Q_j. The threshold is not defined in the standard. If Q_j is below a threshold, the channel j is assumed to be bad. The master then compiles the final list of "good" and "bad" channels to be distributed to every supporting device in the piconet. The master should make the final decision for the channel classification of the piconet.

> In the network there could be legacy devices that are not capable of hopping adaptively. Recall that the hopping sequence is unchanged for all devices at the good channels. This requires the master to transmit broadcast packets only on the good channels. Thus, full support for legacy devices is maintained.

To reduce the time that classification will take, it is possible to reduce the number of measurements at each channel. The procedure is to group channels into blocks and classify the blocks instead of the channels. This will, however, compromise the accuracy of the measurements at each channel.

One-slot packets (such as DM1 or DH1 packets) may also shorten channel classification time. Classification should be performed constantly, both offline and online. Classification when a piconet is not established or the piconet is on hold and no traffic is being exchanged is called offline classification. This classification can involve background RSSI measurements.

PHY and MAC models

The recommended practice relies on analytical models for the physical layer transmitters of IEEE 802.11b and IEEE 802.15.1. Two cases are considered for the simulation results. The first case is the performance of IEEE 802.11b when Bluetooth is considered interference. The second case is the performance of Bluetooth when IEEE 802.11 is considered interference. The model is supplied with device positions and transmission parameters. The path loss model that is used is described by the free-space propagation described by Equation 1–6 with a coefficient 2 up to a distance of 8 m, and coefficient of 3.3 at distances over 8 m. To calculate the BER it can be assumed that an IEEE 802.15.1 device transmits two packets (Figure 3–10). An IEEE 802.11b device transmits a single frame, consisting of a header part and a payload (Figure 3–10). Notice the overlap among the packets. In general, there are six periods of stationarity. A new period of stationarity starts at the end of the IEEE 802.11b header, because the payload is transmitted using a different modulation technique than the header. By definition, during the period of stationarity, the transmit power and modulation type do not change, and the position of the devices is constant. This enables the BER to be calculated.

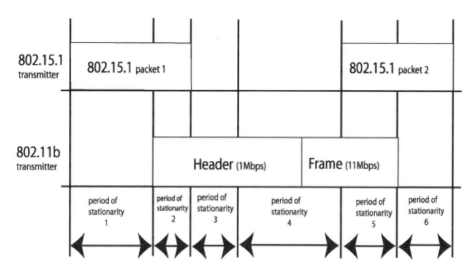

Figure 3–10: An example of interference between IEEE 802.15.1 and IEEE 802.11b with six different periods of stationarity

Figure 3–11 shows a diagram of the BER calculation. The figure shows that the received power of the intended transmission is the EIRP minus the path loss. The interference power at the receiver is the EIRP of the interfering transmission minus the path loss minus an additional factor (called spectrum factor) to account for the combined effect of receiver and transmitter masks. The total interference power is the sum of all interference powers. The SIR is the ratio of the received signal power to the total received interference power. Based on the SIR, first the the Symbol Error Rate is calculated for each modulation type. Then the BER is calculated from the Symbol Error Rate. Note that this model does not consider receiver noise. One simple way to account for receiver noise is simply to decrease the received signal power by a constant, equal to the expected value of the receiver noise. The recommended practice provides analytical formulae for evaluating the BER for the different modulation types in IEEE 802.11 and IEEE 802.15.1.

In addition to an analytical model, the recommended practice includes detailed Monte Carlo simulations. While the analytical model uses transmitter power and distance as input parameters, the simulation model uses signal-to-noise ratio and the signal-to-interference ratio. In both cases, the output is the BER.

To simulate the MAC of IEEE 802.15.1 and IEEE 802.11, the OPNET modeler, a network technology development environment, is used. Each of the IEEE 802.11 and Bluetooth MAC protocols is implemented as a state machine. Events trigger transitions from one state to another. The MAC models were interfaced to the PHY models to simulate a complete system. At the end of each packet transmission, a list is generated consisting of all interfering packets, the collision duration, the timing offset, the frequency, the power, and the topology used. This list is then passed to the PHY module along with a stream of bits representing the packet being transmitted.

For Bluetooth, two types of applications—voice and Internet traffic—are considered. Packet loss measures the number of packets discarded at the MAC layer due to errors in the bit stream. This measure is calculated after performing error correction. The residual number of errors in the Bluetooth voice packets measures the number of errors that remain in the packet payload after error correction is performed.

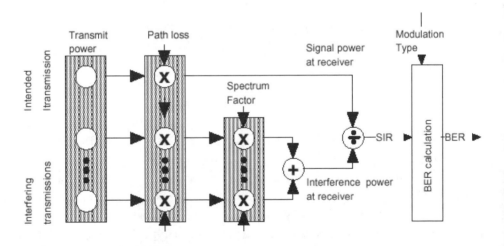

Figure 3–11: BER calculation process

HIGH-RATE WPAN

Overview

As discussed in this book's introduction, a fundamental law of communications is the constant desire to increase the data rate. In November of 1999, before IEEE 802.15.1 was completed, several companies (including Kodak™, Motorola, Cisco, and Aware) initiated the work on a higher-speed WPAN technology, which became IEEE 802.15.3. Future portable consumer electronics devices, such as digital cameras, provided the main motivation. These devices must support multimedia traffic that requires high bit rates. The high-rate WPAN technology must support:

a) ad-hoc connections
b) QoS
c) low cost
d) high speed—up to 55 Mb/s

Because multimedia is an especially important application, IEEE 802.15.3 decided that it would not use the MAC of IEEE 802.15.1, i.e., IEEE 802.15.3 will not only be a physical layer. In this sense, IEEE 802.15 is different from IEEE 802.11, because IEEE 802.11 has only one MAC. The fact that there is a market for a high-rate WPAN shows that Bluetooth is not sufficient. Bluetooth is not fast enough for all applications. Studies have shown that consumers want to transfer large files in less time than possible with Bluetooth. An area where Bluetooth is especially inadequate is video. The second answer comes from the fundamental principles of communications—higher data rate is always desirable, due to a combination of consumer demand and competition among vendors in the marketplace. If Bluetooth is not fast enough, yet another question is why not 802.11b, 802.11a, or 802.11g?. While 802.11 enjoys considerable advantage in the marketplace, as always, when there is a market need, there are several competing technologies trying to satisfy this market need.

Figure 3–12 shows a reference model of IEEE 802.15.3. In addition to a physical layer and a medium-access control layer, as shown in Figure 3–12, both the MAC and PHY layers have management entities, called correspondingly the MAC layer management entity and PHY layer management entity (MLME and PLME, respectively). These management entities interact with the Station Management

Entity (SME). The SME is independent from the physical and MAC layers and is not specified in the standard. In general, the station management entity is responsible for tasks such as the gathering of information from the PHY and MAC layer management entities, and similarly setting the value of certain parameters.

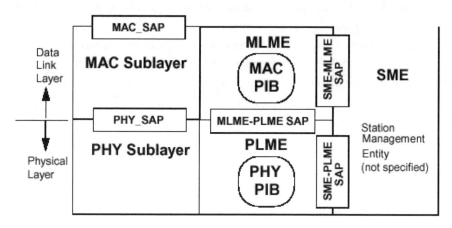

Figure 3–12:   IEEE 802.15.3 Reference model

Figure 3–12 also depicts the relationship among management entities. The physical layer can be considered as providing services to the MAC layer. The five different entities interact via service access points (SAP). The format of these interactions is defined explicitly within 802.15.3. Defined primitives are exchanged over the service access points. The primitives fall into two basic categories: (1) primitives that support peer-to-peer interactions; and (2) primitives that have local significance and support sublayer-to-sublayer interactions. While the format of the primitives is defined, most interactions are not defined explicitly within the standard. For example, the interfaces between the MAC and its management layer, and the PHY and its management layer are not defined. Figure 3–12 is a conceptual diagram. The physical layer can be controlled by the MAC and the MAC management layer. In this case, there is no physical layer management. Vendors may opt to keep the service access points nonobservable. In this case, these vendors can implement these interfaces in any way they wish.

Like the other wireless standards within IEEE 802, IEEE 802.15.3 appears to higher layers starting with the logical link control (LLC), as an IEEE 802-style LAN. However, above the MAC layer, there may not be a 802.2 layer, but a USB or IEEE 1394 convergence layer.

The next section of this chapter is devoted to the MAC layer of IEEE 802.15.3. It is followed by a section devoted to the physical layer.

MAC layer

The IEEE 802.15.3 MAC provides for the following functionality:

1) Allow stations to form and terminate personal area networks.
2) Authenticate stations with each other.
3) Transport data between stations with multiple QoS levels and optional privacy.
4) Allow stations (including the coordinator) to minimize power requirements and still maintain the network.
5) A procedure for the coordinator to pass coordination to another station in the WPAN.

The piconet of IEEE 802.15.3 is shown in Figure 3–13. There are two types of devices: basic devices and piconet coordinators (PNC). One piconet can have at most 255 devices. The PNC (1) provides the basic timing for the piconet with the beacon; (2) manages the QoS; (3) manages power-save modes; and (4) handles the authentication. All devices are not required to be able to act as piconet coordinators. In this way, to achieve a lower cost, some devices do not have full functionality.

Similar to IEEE 802.15.1, the roles of a device and the piconet coordinator can be exchanged on the fly, i.e., a device can take over as a coordinator. This flexibility will be necessary when a coordinating station is removed from the WPAN—for example, when it powers down or moves out of range. Exchanging the roles of the coordinator and a device can take place if the coordinator determines that another station is better suited to perform this role. Note that, similar to IEEE 802.15.1, the coordinator handover requires support at the application layer, beyond the features of the standard. For example, changing the coordinator will usually mean changing the QoS parameters.

A device can start its own piconet or associate with an existing piconet. If a device wants to start its own piconet, it must make sure there is not an IEEE 802.15.3 piconet already started in the same channel. Before starting a piconet, the device will use passive scanning to detect an existing piconet. During passive scanning, the device will look for beacon frames from a PNC. Devices search for the piconet by starting from the first channel and traversing through all the indexed channels available in the PHY. If there are available channels, the device can take the responsibilities of a PNC, choose a channel, and start sending beacons to start its own piconet. If no channels are available, the device can start a dependent piconet.

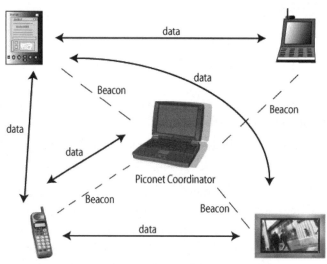

Figure 3–13: The piconet structure of IEEE 802.15.3

Figure 3–13 shows a piconet with no dependent piconets. More complex ad-hoc networks can also be formed, similar to the scatternet architecture of IEEE 802.15.1. A device that participates in a piconet can form a dependent piconet. The dependent piconet depends on time allocation from the parent piconet and is synchronized to the parent piconet. The two types of dependent piconets are child and neighbor piconets. A parent piconet can have more than one child piconet, and a child piconet, in turn, can have a child or neighbor piconet of its own. One piconet can have simultaneously more than one child and neighbor piconets.

The child and neighbor piconets are autonomous piconets and have distinct piconet identifiers (PiconetIDs). They depend on a dedicated time slot called channel time assignment (CTA) from the parent piconet. The coordinator of the child piconet is a device in the parent piconet. Therefore, the child piconet extends the area of coverage of the main piconet. A parent/child piconet combination is architecturally just like a IEEE 802.15.1 scatternet. A parent/child piconet combination poses the question of efficient routing. (This question has not been studied and is a very suitable topic for future research.) The coordinator of the neighbor piconet is not participating in the parent piconet. Therefore, the dependent piconet does not extend the area of coverage of the parent piconet. Dependent piconets are normally used to create another piconet even when no channels are available. Following a time-division multiplexing principle, dependent piconets work in the same frequency spectrum as the independent piconet. The superframe of the dependent piconet fits in one guaranteed time slot of the parent piconet. A device that wants to form a child piconet of its own will send a request for a CTA to its PNC.

Piconets are discovered by passive scanning, in which devices listen for a beacon. A station desiring to join an IEEE 802.15.3 WPAN will set its receiver to periodically listen on the various channels for a beacon. If the device finds a beacon, and the beacon indicates a network of interest, it will attempt to authenticate with the coordinator. Security in all standards belonging to the IEEE 802.15 family is provided on the assumption that there is no connection to an external network. This is an important difference between IEEE 802.15 and IEEE 802.11. In addition, generally IEEE 802.15 devices have less computational power and less memory. It is also assumed that an attacker can have more computational power and more memory. There are two security modes—mode 0 and mode 1. In mode 0, authentication and encryption are not required. The only security service available in mode 0 is an access control list (ACL). The security services provided in mode 1 are access control list, mutual authentication, key establishment, key transport, and verifying the authenticity of keys, data encryption, data integrity protection, beacon integrity protection, freshness protection, and command integrity protection.

Stations wishing to join the WPAN authenticate with the coordinator and then with any other station with which communication is required. This authentication can be "open," i.e., only the public key information of the station to which authentication is attempted is used. It may also require that a secret PIN is known and transferred during the exchange, thus allowing controlled authentication. A secret key is generated to allow privacy for any subsequent data exchange. Upon successful authentication, the station is considered to be in the WPAN. It will also exchange capability information with the coordinator. This capability information includes all the PHY data rates supported by the station, its power management status, whether the station can be a coordinator, buffer space, etc. As a result of this exchange, a coordinator handover may occur. All the frames between the PNC and the device before the completion of the association of the device are exchanged in the contention access period only. To participate in a piconet, a device must join by association. The unregistered device initiates the association process by sending an association request. This association request is sent during the contention access period. When the PNC receives a valid association request, it sends an acknowledgment first, and after some time sends the association response, indicating acceptance or rejection of the request. If the association is successful, the coordinator informs the new device about the services offered by the piconet, and the new device informs the coordinator of the services that it offers. As a result the beacon is changed to include information about the new device, so that the rest of the devices are also informed about the services that the new device offers. In a very similar way, when a device leaves the piconet, the information about this device is removed from the beacons.

Security mode 0 is mandatory to implement. Security mode 1 is optional. Mode 1 offers three security suites: elliptic curve Menezes-Qu-Vanstone (ECMQV) Koblitz-283, NTRUEncrypt 251-1, and RSA-OAEP 1024-1. The ECMQV security suite has three subsuites: ECMQV manual that does not accept certificates, ECMQV implicit, and ECMQV X.509 that accept certificates. RSA-OAEP 1024-1 has two subsuites, one that does not accept certificates and another that does. Devices that support mode 1 must implement at least one security subsuite. The ECMQV Koblitz-283 security suite provides 128-bit security, the NTRUEncrypt 251-1 security suite provides 80-bit security, and the RSA-OAEP 1024-1 security suite provides 80-bit security.

All security suites support CTR + CBC – MAC (CCM) combined encryption and authentication. The CCM mechanism consists of the generation of an integrity code followed by the encryption of plaintext data and the integrity code. The output consists of the encrypted data and the encrypted integrity code. The symmetric authentication operation used in this security suite consists of the generation of an integrity code, using a block cipher in CBC mode, computed on a nonce, followed by (optional) padded authentication data, followed by (optional) padded plaintext data. The verification operation consists of the computation of this integrity code and a comparison to the received integrity code. The symmetric encryption operation used in this security suite consists of the generation of a key stream (using a block cipher in counter mode with a given key) and nonce, and performing an XOR of the key stream with the plaintext and integrity code. The decryption operation consists of the generation of the key stream and the XOR of the key stream with the cipher-text to obtain the plaintext and integrity code. CCM is parameterized by the use of the Advanced Encryption Standard (AES) with the length field L set to 2 bytes, the length of the authentication field M set to 8 bytes, and the length of the nonce set to 13 bytes. AES is 128 bit, performed as specified in NIST FIPS Pub 197 [B113].

The ECMQV Koblitz-283 security suite uses a cryptographic hash function. All operations are on elliptic curves. An elliptic curve key pair consists of an integer q and a point Q on the curve determined by multiplying the generating point G of the curve by this integer, i.e., $Q = qG$. Here, Q is called the public key, and q is called the private key. The ECMQV algorithm can use different types of certificates. The manual certificate is a concatenation of the public key and the device address. The manual certificate is not a digital certificate, because the binding between a public key and its purported owner is to be

> An implicit, manual, or digital ECC certificate is verified by first extracting the public key and then extracting the identifying string of its purported owner. Then, the identifying string U of the purported owner is compared with the device address.

established and verified by noncryptographic means. The implicit certificate scheme used with the ECMQV security suite is the implicit certificate scheme with associated implicit certificate generation protocol and implicit certificate processing protocol ANSI X9.63-2001. The explicit certificate scheme used with the ECMQV security suite is the ECC X509 certificate scheme, with associated ECC X509 certificate generation protocol and ECC X509[10] certificate processing protocol, as specified in IETF documents RFC 2459 [B85], RFC 3279 [B86], and RFC 3280 [B87]. The certificate is generated and verified using the digital signature algorithm and verification protocol ECDSA, which is fully specified in ANSI X9.62-1998.

The NTRUEncrypt 251-1 security suite uses a cryptographic hash algorithm as specified in the FIPS 180-2 draft standard [B44]. This security suite has a certificate-based security subsuite. The digital certificate format and verification protocol are as in [B87].

The RSA-OAEP 1024-1 security suite is based on modular exponentiation. An RSA public key in this security suite consists of the modulus n and the public exponent e. The private key is information that allows its owner to find e-th roots mod n. The raw public key is the modulus concatenated with the exponent. The encryption scheme in this security suite is as specified in [B118].

Authentication is mutual and based on public-key cryptography. However, to achieve mutual authentication with this system, the devices must trust that the key is indeed associated with the intended device. This trust can be achieved in several ways. One way is to use access control lists (ACLs). The ACL consists of device addresses and other device-specific information such as keys. Before the authentication, the ACL can have information about the key and its associated device ID. For example, some vendors can preload their products with appropriate ACLs to facilitate authentication with certain other devices. Another way to establish trust is by the verification of digital certificates at the time of authentication. Yet another approach is by user action, where, for example, the user pushes buttons to facilitate the authentication process. All of these processes are outside of the 802.15.3 standard.

As a result of the authentication, shared keys are established. Data encryption is performed by a symmetric cipher. Data integrity and sequential freshness are other security services. The beacon may be integrity-protected to assure the devices that indeed, the beacon comes from the coordinator. To maintain security, any device that has performed mutual authentication with another device, can request that the other device periodically transmit a secure frame to indicate its continued presence in the piconet. The standard also ensures that only current

---

[10] X.509 is part of the ITU X.500 series of recommendations. It specifies the form of the authentication information, and the way to use this information to perform authentication. The certificate can be freely communicated. It contains a user's public key that has been signed by a certificate authority.

members of the piconet can transmit and receive data. Therefore, when a device joins or leaves the piconet, the group keys are changed.

Medium access mechanism

To support QoS efficiently and to maintain the very successful 802-style CSMA protocol, after considerable deliberation and compromise, IEEE 802.15.3 decided to make the medium access mechanism a combination of CSMA and TDMA. In this way, the advantages of both access mechanisms are combined, and the MAC can efficiently transport asynchronous and isochronous data. The MAC superframe structure consists of three parts: a network beacon interval, a contention access period (CAP), and a contention-free period (CFP), as shown in Figure 3–14. The minimum superframe duration is 512 µs and the maximum superframe duration is 65,535 µs. The beacon contains information about the duration of the superframe, the duration of the CAP, and the contention-free period. The beacon also has information about the identity of the network so that devices may join.

The beacon is also used to synchronize the network coordinator and the devices. Because the medium access mechanism is a combination of CSMA and TDMA, synchronization is very important. The network coordinator contains the reference clock. The beacon sent at the beginning of every superframe contains as a time-stamp the value of the coordinator timer. This time stamp value is equal to the value of coordinator's local timer value at the time of transmission of the first bit of the time-stamp field. Because the frame checking is done after receiving the entire frame, all devices make a copy of the local timer and keep the copy until the end of the beacon frame and adjust it to make it equal to the clock of the coordinator. Unassociated devices that wish to join the piconet synchronize using the same mechanism.

A device must hear at least one beacon before sending its request for association. All devices resynchronize their clocks with the coordinator when they receive the beacon. In addition, the child and neighbor piconet coordinators synchronize their clocks with the coordinator of the parent piconet. The superframe duration can be changed at the discretion of the network coordinator. The contention-free period is reserved for channel time allocations (CTA), including management CTA (MCTA). The boundary between the CAP and CFP periods is dynamically adjustable. The CAP period is reserved for transmitting non-QoS data frames. The medium-access mechanism during the CAP period is CSMA/CA. The remaining duration of the superframe is reserved for CTA to carry data frames with specific QoS provisions. Finally, in power-save mode, all QoS features are maintained.

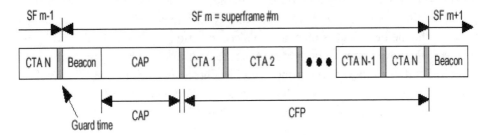

Figure 3–14:   Superframe of a IEEE 802.15.3

Four interframe spaces (IFS) are defined: short (SIFS), minimum (MIFS), backoff (BIFS), and retransmission (RIFS). The actual values of these interframe spaces are PHY dependent. Both in the CAP and the CFP, the acknowledgment starts within SIFS after the end of the transmission of the previous frame. Similarly, SIFS duration is allowed between a frame that does not expect an immediate response and the next successive frame. During the CFP, all devices use RIFS for retransmissions. The beacon is transmitted RIFS after the previous transmission. When acknowledgment is not required, in an CTA during the CFP, the transmission of two successive frames will be separated by MIFS.

All data is exchanged in a peer-to-peer manner. Allocating a one-time CTA is similar to polling. The device requests from the coordinator the total amount of time necessary for communication. The PNC grants or rejects this one-time request. Unlike IEEE 802.15.1 all traffic during both the CAP and the CFP is exchanged between peer devices. This ultimately improves latency and the QoS.

It should be noted that the CSMA/CA mechanism in the CAP is slightly different from the one used in IEEE 802.11. In particular, during the CAP no

> There are three methods of communication. First, a device can contend for the medium during the CAP using the CSMA/CA scheme. Second, the device can have allocated CTA on a periodic basis for periodic traffic, and third, the CTA may not be on a periodic basis.

transmission can extend past the duration of the CAP. A device is not allowed to extend its transmissions that started during the CAP in to CFP. Hence, once device decrements its backoff counter to zero, it will check whether there is enough time in the CAP for the transmission of current frame and SIFS. If there is not enough room for this entire frame exchange sequence, then the device will abort the transmission and choose a backoff value with minimum allowed backoff-window size.

A backoff value is randomly (uniformly distributed) chosen in a certain range. The retry count is zero for the first transmission attempt. A backoff counter is maintained, and it is decremented only when the medium is idle for the entire interval minus the RX-TX turnaround time. Whenever the channel is busy, the backoff counter is suspended. When the backoff counter reaches zero, the device will transmit its frame. The backoff counter is also suspended outside CAP duration. In this way, the backoff counter is maintained across multiple superframes and is not reset with every beacon. Immediate acknowledgment is sent without the backoff procedure. The retransmission and the timeout are the same as in IEEE 802.11. The backoff is exponential, where the size of the backoff window is doubled after every unsuccessful transmission attempt. Because the beacon is used for synchronization, the backoff procedure is not applied to the transmission of the beacon. During CAP, a device is allowed to transmit one frame at a time with backoff being applied to every frame. After transmission, if the device does not receive the expected immediate response (acknowledgment), the retry count is incremented. If the maximum number of retries has not been exceeded, the backoff procedure is again resumed. Instead of DIFS, as in IEEE 802.11, a device first waits for a duration equal to BIFS before beginning the backoff procedure.

The PNC divides the CFP into several guaranteed time slots (GTS). Each GTS is a single time slot with guaranteed start time and a guaranteed time duration reserved within the CFP. Hence, a device that is allocated with a GTS is guaranteed that no other devices will compete for the channel during the indicated time and duration of the GTS. During the CFP channel access is entirely controlled by the network coordinator. The PNC assigns channel time allocations to individual devices in a TDMA fashion. The PNC allocates these guaranteed time slots based on the currently pending requests by all the devices in the piconet and the available channel time within CFP. If a device needs channel time during the contention-free period, it makes a request to the PNC. These requests take place during the contention access period. There are two types of requests—requests for isochronous or periodic channel time and requests for aperiodic channel time. Rather than requesting recurring channel time, an asynchronous channel time request asks for a total amount of time. The coordinator evaluates the current usage for each new request for QoS and determines if the service can be granted. If the resources are available, the PNC allocates time in a GTS for the device. Note that the allocation by the PNC may not exactly match the duration of time requested by the device. If the requirements for the data change, then the device can request a change to the allocation. The allocated CTA can be shortened or terminated by the transmitting device or the PNC. Upon grant from the coordinator, the device will begin using the CFP in the *next* superframe. The coordinator will assign a stream index to the requesting station. Because all communication is between peer devices, CTAs are specified by a source address and a destination address. Devices cannot extend their transmissions beyond their allocated CTA, including an immediate acknowledgment (if required).

Special CTAs for commands are called management CTAs. They are distinguished by the source address and destination address being the PNC. Adjacent CTAs are separated by guard times. In addition, after the last transmission in a CTA and the beginning of the guard period, there is time period equal to SIFS.

A device with a GTS may or may not make use of all the allocated time duration within the GTS. What to transmit during a GTS is determined locally by the device depending on priority. The devices associated with a PNC can send their changes in channel time requirement. The slot assignments within CFP are based on the current pending requests from all the devices and the currently available channel time within CFP. The slot assignments may change from superframe to superframe as felt required by the PNC.

In no case can a device extend its transmissions beyond the end of the time slot where transmission began. Therefore, the source device will always check whether there is enough time in the time slot for the transmission of the current frame plus an amount of time equal to SIFS. In addition, if an immediate acknowledgment is expected for that frame, the remaining time in the time slot must be large enough to accommodate the current frame, the immediate acknowledgment, and an amount of time equal to twice SIFS. If there is not enough room for this entire frame exchange sequence, then this device must abort the transmission and leave the remaining duration of the time slot unused. Within the CFP, however, when an acknowledgment is expected but not received within RIFS, the source device will retransmit the frame as long as there is enough time during its CTA.

Information about the CTAs is included in the beacon. However, due to characteristics of the wireless environment, in any superframe there may be one or more devices in the piconet that receives the beacon in error, depending on their location and interference. If a device misses the beacon, it will not transmit in its CTA. To maintain QoS, even if a device misses the beacon for several superframes, it may still transmit in its CTA, if the CTA is pseudostatic. If a device misses the beacon for longer periods, it will assume that the piconet no longer exists.

Acknowledgment and retransmission

There are three types of acknowledgments: immediate, delayed, implied, and in addition, there is simply no acknowledgment. A transmitted frame with an acknowledgment policy set to indicate no acknowledgment is not acknowledged. The transmitting device assumes that the transmission has been successful. Broadcast and multicast frames are some of the frames that do not require acknowledgments. If the policy is immediate acknowledgment, then following a correct receipt of a frame, the recipient starts the transmission of the acknowledgment within a time period equal to the SIFS. Delayed acknowledgment allows the receiver to group the acknowledgment into retransmission request command. Delayed acknowledgments can be sent anytime before the expiry of the retransmission window. The transmitter can solicit a delayed acknowledgment. Delayed acknowledgment should be used only for periodic traffic. The third type of acknowledgment, implied acknowledgment, is applicable only during the contention-free period. Implied acknowledgment is not used during the CAP. Implied acknowledgment allows the receiver to send another data packet in response. This response data frame is sent after a SIFS period, just like immediate acknowledgment. In turn, the response data frame can set its own acknowledgment policy, which can be of any type except implied.

During the CAP, retransmissions (up to a certain maximum number) are performed according to the backoff rules. During the time slots within the contention-free period when either immediate acknowledgment or implied acknowledgment is expected, the transmitting device performs CCA between the end of SIFS and the end of RIFS to detect the start of the acknowledgment frame. If an acknowledgment is not detected, the source device can start the retransmission of the frame (or new transmission) at the end of RIFS as long as there is enough channel time remaining in the time slot for the entire frame exchange. When an immediate acknowledgment is expected and received, the transmitting device waits for the duration of immediate acknowledgment frame plus SIFS before attempting another transmission. When an implied acknowledgment is expected and received, the source device waits for at least the duration that was indicated in its immediately previous transmission or the CCA indicating that the channel is idle.

Network Robustness

IEEE 802.15.3 includes measures for network robustness. These measures are transmitter power control, dynamic channel selection, and network coordinator handover.

Dynamic channel selection and transmitter power control are parts of 802.15.3. Two types of transmitter power control are available. During the CAP, the power level is fixed. This prevents some devices from unfairly gaining access to the wireless medium, due to their higher transmit power. During the CFP, however, the power levels are adjustable. The two devices that communicate during a CTA can determine the power level to be used.

This technology also supports dynamic channel selection. To overcome the problems due to overlapping piconets and interference in a given channel, the PNC may choose to move the operations of the piconet to a new channel. Once the coordinator has started a piconet, it will periodically allocate channel time in the CFP to scan the channel for other piconets. This considerably improves the coexistence performance. If the coordinator finds another piconet, it can become the child or neighbor piconet of the other piconet, or it can change to a different channel.

The network coordinator monitors channel conditions by performing passive scanning or gathering information from the other devices. If it decides that channel conditions are not appropriate, it can dynamically change the channel the piconet is operating on without any disruption in services. The PNC collects the channel status from the devices to arrive at this decision. There is a channel status request command to request that the devices inform the PNC of their channel status. The algorithm to use the channel status information and decide how to perform the change of channel is beyond the scope of 802.15.3.

If the decision is made by the PNC to change the channel, the PNC keeps the piconet quiet by not transmitting any beacon for one or more beacon intervals. Within that quiet time, the PNC may change to one or more other channels to check if one of the other channels is better than the current channel, and then return back to the current channel. If no beacons are received for a certain duration, the devices assume that they are disconnected and restart the association process. If the PNC returns to the current channel within the indicated timeout, the PNC sends a beacon to cancel the quiet state of the piconet. That beacon can include the channel change element indicating the new channel and the timeout for changing channels. The devices that received the beacon with the channel change element will change the channel to the indicated new channel within the indicated time period. They will start waiting for beacons in the new channel. If another channel is not available, the PNC may decide to stay in the same channel. In this case, it will not send a channel change element in its beacon following the cancellation of quiet state of the piconet.

The devices in the piconet can use power-saving techniques to reduce their power consumption. Each device in the piconet can choose transmission power based on the current channel conditions. Furthermore, stations not using a data stream can reduce their power requirements by turning off their physical layer until the next superframe begins. Stations need only be awake during the beacon and the subsequent RTS window of the CAP. If devices are known to have their PHY inactive, the RTS window is used to notify the station of a message. If no message for a station is indicated in the RTS window, and the station does not have activity for the CFP, it can power down until the next beacon. The power state of a device is communicated to other devices by streamless data exchanges. These streamless data exchanges take place during the contention access period. Data can be exchanged in an allocated stream in the CFP, or for small amounts of data without

QoS requirements, in the data window of the CAP. In the second case, if the destination station is not known to be powered up, the RTS window is also used, and the data will be exchanged only if the destination responds.

Every device can be in one of two states—awake or sleep. Devices that transmit or receive are in the active state. There are four modes: active, hibernate, piconet synchronized power save (PSPS) and synchronous power save (SPS). Devices that are in the hibernate, PSPS, or SPS modes can be awakened by a awake beacon. The awake beacon for devices in the PSPS and SPS modes is periodic. In SPS mode, several devices form a group and can be awakened simultaneously. In hibernate mode, devices conserve power and do not listen for beacons. These modes alone are not enough for battery-powered devices. In addition, these devices shut down during parts of the superframe when they neither transmit nor receive. For example, devices will shut down during the GTS of other devices.

> Note that there is a device information request command that each device can use to obtain the information about other devices in the piconet. In this way, peer discovery and peer capability discovery are performed. Thus, the PNC knows the capabilities of the devices in the network.

If the current piconet coordinator is about to be powered down or is moving out of range, it must attempt to choose a device that is capable of being a PNC as its successor. The PNC will send a coordination-handover command to its chosen device with an indication of the handover timeout. The device must accept the nomination and obtain the device information from the current PNC within the indicated timeout period. The new PNC must announce its new responsibility as PNC by sending its first beacon at the first expected beacon transmission time.

When the coordination handover is successful, the association of the remaining devices with the piconet is unaffected, and hence they are not required to reassociate with the new PNC. Note that the coordinator handover need not always necessarily stop all the stream transmission.

Each MAC frame consists of a header, a variable-length frame body, and a frame-check sequence (FCS). The header includes a header check sequence, which is verified by the PHY, i.e., the PHY passes on to the MAC frames, which pass the HCS test. The FCS is a 32-bit cyclic redundancy code (CRC). The CRC is calculated over the entire frame body. The maximum size of a MAC frame is 2048 bytes and the minimum frame body is zero bytes. Fragmentation is performed at the transmitting device on all frames whose size is greater than a certain fragmentation threshold. All fragments except the last must be of the same size. There is no theoretical limit for the size of frames with this fragmentation mechanism.

Physical layer for the 2.4 GHz ISM Band

At present the IEEE 802.15.3 physical layer operates in the 2.4 GHz band. Other physical layers are currently under development. The symbol rate is 11 M symbols per second. Six distinct modulation formats are supported at this symbol rate: uncoded DQPSK, QPSK, trellis-coded QPSK, and 16/32/64-QAM. To achieve low-cost and low-power consumption, OFDM was rejected and IEEE 802.15.3 settled on a single-carrier physical layer. The IEEE 802.15.3 signals occupy a bandwidth of 15 MHz. The total number of channels is five, but they can be used in two different ways, or modes. The first is the high-density mode, which allocates four channels centered at 2.412 GHz, 2.428 GHz, 2.445 GHz, and 2.461 GHz. The second is an IEEE 802.11b coexistence mode, which allocates three channels centered at 2.412 GHz, 2.437 GHz, and 2.461 GHz. A device that is scanning the spectrum for beacons will scan all channels. Furthermore, a device that wants to start a piconet can scan the IEEE 802.11b channels for IEEE 802.11b networks, if it is capable of detecting IEEE 802.11b. If a coordinator detects an IEEE 802.11b network, it will use the IEEE 802.11b coexistence mode.

If a device supports a given QAM modulation format, it also must support all of the lower modulation formats. The base rate of the IEEE 802.15.3 2.4 GHz PHY is 22 Mb/s, uncoded DQPSK mode. DQPSK is used instead of the BPSK to reduce the overhead due to the duration of the PHY and MAC headers. Differential encoding also allows noncoherent receivers to be used. A device can send a frame with one of the supported data rates to a destination device only when the destination device is known to support that rate. All group-addressed frames, regardless of their type, are sent at the base rate that is supported by the piconet because all devices must be able to receive these frames.

Figure 3–15 illustrates the signal constellations used in encoding bit streams into discrete signal levels sent through the air medium. The average power within a packet including the PHY header is required to be a constant, regardless of the modulation format. Thus, the constellation must be scaled by a normalization factor such that the PHY header and the frame body have the same average power.

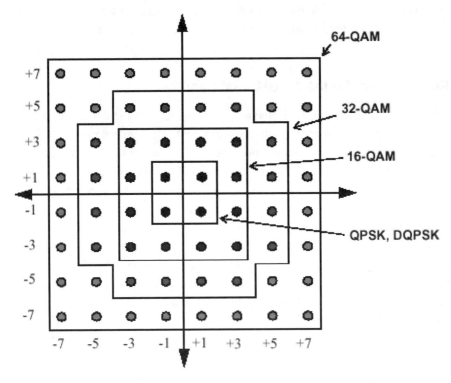

Figure 3–15: DQPSK, QPSK, and QAM constellation diagram

The higher data rates of 33–55 Mb/s are achieved by using 16, 32, or 64-QAM schemes with trellis coding. Coded modulation schemes such as trellis coding are known to provide better coding performance [B134]. In trellis coding, not all information bits are encoded, which results in an overall coding gain. The trellis code that is applied to QPSK and 16/32/64 QAM is a 2-D eight-state (2D-8S) trellis code, as shown in Figure 3–16. The constellation points of the modulation are partitioned into eight subsets: D0, D1, D2, ... D7. Each subset contains two, four, and eight symbols for 16/32/64-QAM, respectively. According to this trellis code for 16/32/64 QAM, the lower-order three bits select one of these eight subsets. The remaining higher-order bits select one of the symbols from each subset. In the trellis-coded QPSK mode, the lower-order two bits select one of the four subsets D0, ... D3, each containing a single symbol.

For 64-QAM, five input bits per symbol are encoded into six output bits. For 32-QAM, in a similar fashion, four input bits per symbol are encoded into five output bits, and for the case of 16-QAM symbol mapping, three input bits per symbol interval are encoded into four output bits. Finally, as shown in Figure 3–16, a 16/32/64 QAM constellation point is selected.

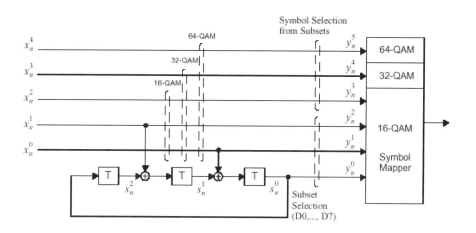

Figure 3–16: Trellis coding diagram for 16/32/64 QAM

For a given output bit vector of the trellis encoder, a 16/32/64-QAM constellation point is selected based on the set partitioning rule illustrated in Figure 3–17 and for QPSK as illustrated in Figure 3–18. These figures also show the bit-to-symbol mappings. The lower (rightmost on the figures) three bits correspond to the subset numbers. The higher 3 bits are assigned within each signal subset (D0, ...D7) such that decimal representation of the bit mapping goes from low to high as the constellation points are traced from bottom to top and from left to right. This rule ensures that the decimal representations of the bit mappings from 0–15 belong to 16-QAM, and 0–31 belong to 32-QAM, and 0–63 belong to 64-QAM.

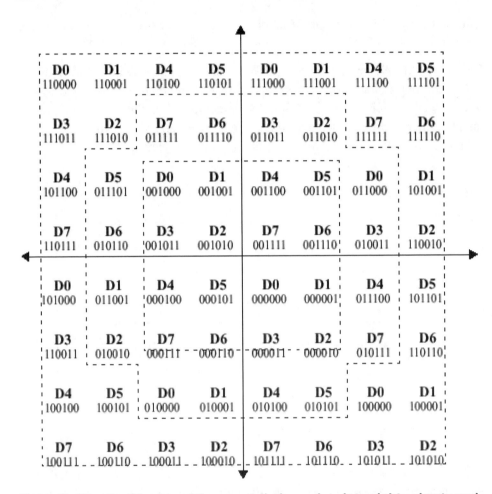

Figure 3–17: Partitioning of the constellation points into eight subsets and bit-to-symbol mapping for 16/32/64 QAM

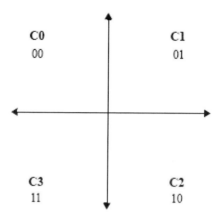

Figure 3-18: Bit-to-symbol mapping and set partitioning for QPSK

The differential encoding applies only to the QPSK mode. In this mode, the entire frame, with the exception of the PHY preamble, is encoded differentially. The differential encoding is provided to allow for noncoherent and, therefore, lower cost receiver implementations.

The IEEE 802.15.3 physical layer adds a PHY header to the MAC header, and calculates the header check sequence (HCS) over the combined MAC and PHY headers. The PHY header has information about the length and the data rate of the payload. The HCS is appended to the end of the MAC header. While the maximum MAC frame is 2048 bytes, this includes the frame body and FCS but not the PHY preamble, PHY header, MAC header, or HCS. The header check sequence contains the 16-bit CRC for the combined PHY and MAC headers. This 16-bit CRC is the 1's complement of the remainder generated by the modulo-2 division of the combined PHY and MAC headers by the polynomial $x^{16} + x^{12} + x^5 + 1$, as in the CCITT CRC-16, and the same as in IEEE 802.11b.

A frame sent over the air is illustrated in Figure 3-19 for data rates of 22, 33, 44, and 55 Mb/s, and in Figure 3-20 for the data rate of 11 Mb/s.

At the rate of 11 Mb/s, two repetitions of the PHY header—MAC header and HCS—are transmitted, as shown in Figure 3-20. The first repetition is sent using 22 Mb/s DQPSK. The second repetition is sent using 11 Mb/s QPSK + TCM. This repetition makes the header more robust, which is desirable because the header is more important.

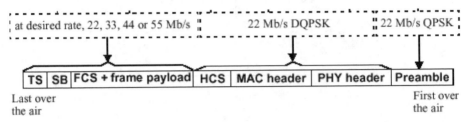

Figure 3–19: Transmitted frame for data rates of 22, 33, 44, and 55 Mb/s

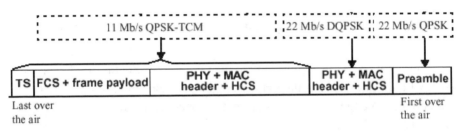

Figure 3–20: Transmitted frame at 11 Mb/s

A PHY preamble is added to aid receiver algorithms related to synchronization, carrier offset estimation, and channel estimation and equalization. The preamble consists of successively appending 10 periods of a special sequence. The sequence is called constant-amplitude zero autocorrelation (CAZAC) sequence. Each period of the CAZAC sequence contains 16 complex symbols, C_0, $C_1,...C_{15}$, where

$C_i = 1 + j$, $i = 0,...,3$, and $i = 7, 9, 11, 15$

$C_i = -1 + j$, $i = 4, 14$

$C_i = -1 - j$, $i = 5, 8, 10, 13$

$C_i = 1 - j$, $i = 6, 12$

Clearly these complex symbols are of QPSK type. The first nine periods of the preamble are a concatenation of the CAZAC sequence. In the tenth period, each element of the CAZAC sequence is negated, or equivalently, rotated by 180 degrees to denote end of preamble.

The MAC header, HCS, and payload are scrambled, while the PHY preamble and PHY header are not scrambled. The scrambler is described by the polynomial $1 + x^{14} + x^{15}$. There are four initialization vectors for the scrambler. For the first frame, the scrambler is initialized with 001111111111111, for the second, 011111111111111, for the third, 101111111111111, for the fourth, 111111111111111. After the fourth frame the initialization pattern periodically repeats. The first two bits of the initialization vector are sent to help the receiver estimate the initialization vector.

If the number of payload bits does not correspond to integer number of symbols, then stuff bits (SB) are added. The data rate of 11 Mb/s corresponds to one bit per symbol, and therefore stuff bits will not be needed. The use of TCM requires tail symbols. These tail symbols are used to terminate the encoded trellis sequence in a known state to aid the decoding process. At the end of each frame, the trellis should be terminated in state 0. For 11 Mb/s QPSK-TCM format 3 symbols, each containing 1 bit, are appended to the end of the MAC frame body. For 16/32/64 QAM modulation two tail symbols are appended to the end of the payload to terminate the encoded trellis sequence in state 0.

The receiver sensitivity is defined as the minimum received signal level in dBm that guarantees a packet error rate of less than 8 percent. For the different modulation schemes, receiver sensitivities range from –68 dBm to –82 dBm.

For the maximum transmit power, IEEE 802.15.3 devices must satisfy FCC regulations in 15.249.

LOW-RATE WPAN

Overview and architecture

After the development of Bluetooth, it became clear that the one technology-fits-all approach is not appropriate. Many industrial, agricultural, medical, and vehicular applications, such as sensors, meter reading, smart tags/badges, and home automation, require short-range wireless connectivity that is different from Bluetooth. Two parameters are very important for these applications—ultra low power consumption and ultra low cost. Power consumption must be low so that batteries can last several months or longer. Data rate is not that important for these applications, and they are willing to trade it for very long battery life. Bluetooth or the high-rate WPAN technology cannot address these applications. IEEE 802 addressed these market needs by developing low-rate WPAN technology. The standard is called IEEE 802.15.4 and (similarly to the other 802 standards) defines the physical and medium access control layers. In addition to defining these two layers, the standard aims to achieve a certain level of coexistence with other wireless devices.

> The data rate may be low, but the QoS requirements ensure that wireless devices built using this technology can carry signals such as voice.

In general, low-rate WPAN devices can be fixed, portable, and/or moving. The main objective of the 802.15.4 standard is a simple, but flexible and efficient, protocol. It was decided that a IEEE 802.15.4 network is composed of two different device types—a full-function device and a reduced-function device. The full-function device can operate in three modes—as a network coordinator (controlling the network), a coordinator, or a network device. It can dynamically switch its mode of operation. Reduced-function devices cannot control the network and can be extremely simple, which is appropriate for devices such as a light switch or a passive infrared sensor. Reduced-function devices cannot communicate with other reduced-function devices; they can only communicate with full-function devices. As a result, the protocol for reduced function devices can be implemented using minimal resources and memory capacity.

Low-rate WPAN devices achieve data rates of 250 Kb/s, 40 Kb/s, and 20 Kb/s, use CSMA/CA channel access, have power management, and use 16 channels in the 2.4 GHz ISM band. The remainder of this section is devoted to the network topologies for low-rate WPAN, the protocol architecture, and an overview of the constituent protocols.

There are two basic network architectures for low-rate WPAN—star and peer-to-peer, as shown in Figure 3–21. In the star topology, the network nodes can only communicate with the single central controller, i.e., the network coordinator. A network device has some associated function and is either the initiation point or the termination point for network communication. A network coordinator can be used to initiate, terminate, or route communication around the network. The network coordinator does not need to be a host PC. One difference compared with other technologies is that while the nodes are generally bidirectional, unidirectional nodes are permitted with extremely limited network access. Messages between two network devices can be exchanged through virtual peer-to-peer links, where the message frames are routed though the network coordinator. The star topology is appropriate for PC peripherals and home automation products. In addition to the two basic network types, more complex network architectures are possible, such as cluster-tree networks. Therefore, 802.15.4 supports the architectural requirements of wireless sensor networks.

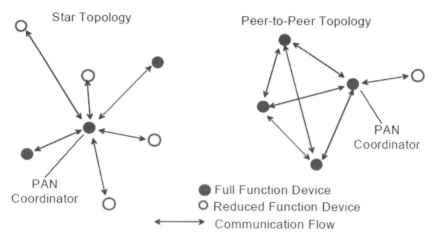

Figure 3–21: Topology of a low-rate WPAN

In the peer-to-peer topology, any node can communicate with any other nodes as long as they are in range of one another. Therefore, the peer-to-peer topology requires only full-function devices. In this mode, more complex network topologies, such as the cluster tree topology, are possible. Applications such as intelligent agriculture, industrial control and monitoring, wireless sensor networks, asset and inventory tracking, and security would benefit from such network topology. Its advantage is that it is ad hoc, self-organizing, and self-healing. It allows multihops to route messages from any node on the network to any other node on the network. While 802.15.4 allows this type of network, the formation and management of a cluster-tree network are performed by a layer above the MAC layer of 802 standards. Furthermore, both the star and

Chapter 3: Standards for Wireless Personal Area Networking (WPAN)

peer-to-peer architectures can be connected to other networks. For example, if the PAN coordinator is a computer, the computer can be connected to other wireless or wired networks.

The LR-WPAN standard consists of MAC and a PHY. A LR-WPAN device is comprised of a physical layer, which contains the RF transceiver and any low-level control of the transceiver, and a medium access control sublayer. The layers interface by one or more service access points.

Power consumption is paramount in this technology. In many applications that use IEEE 802.15.4, the devices will be battery powered. Reduced power consumption is achieved in two ways. First, the IEEE 802.15.4 protocol is developed so that devices require little power to operate. Second, in the physical implementation of 802.15.4, additional power management techniques can be used.

The additional power management techniques are one area of vendor differentiation in the marketplace. Note that some devices could potentially be mains powered. Battery-powered nodes will typically use duty-cycling to reduce power consumption. In fact, these devices will spend most of their existence in a sleep state, not participating in the network. However, each network device must periodically listen to the RF channel for a network beacon in order to determine whether a message is pending from a network coordinator or another device, and to stay synchronized to the network. This mechanism allows the application designer to decide on the balance between battery consumption and message delay latency. Mains-powered network devices have the option to listen to the RF channel continuously.

PHY Layer

The features of the PHY layer are activation and deactivation of the radio transceiver, channel adjustment, and clear channel assessment. To achieve low cost, the radio of the on-air protocol stack operates in the 2.4 GHz ISM band, available worldwide, and in lower ISM bands such as the 868 MHz band, available in Europe, and the 915 MHz band available in the U.S. (Table 3–4). A total of 27 channels, numbered 0–26, are available in these three frequency bands. Sixteen channels are available in the 2.4 GHz band, 10 in the 915 MHz band, and one in the 868 MHz band.

The center frequencies of these channels are:

f_c = 868.3 MHz in the 868–868.6 MHz band, available in Europe,

f_c = 906 + 2(k − 1) MHz, k = 1,…,10, in the 900 MHz ISM,

f_c = 2405 + 5(k − 11) MHz, k = 11,…,26, in the 2400 MHz band.

The physical layer in the 2.4 GHz band uses 16-ary quasi-orthogonal modulation. Four information bits select one of the 16 nearly orthogonal PN sequences to be transmitted. The complex chip sequence is modulated on the RF carrier using offset-QPSK modulation. Even-indexed chips are modulated onto the I carrier, and odd-indexed chips are modulated onto the Q carrier. According to offset QPSK, the Q-carrier will be delayed with respect to the I-carrier by half of the chip period. Each PN sequence has 32 chips. These 16 sequences are:

```
1 1 0 1 1 0 0 1 1 1 0 0 0 0 1 1 0 1 0 1 0 0 1 0 0 0 1 0 1 1 1 0
1 1 1 0 1 1 0 1 1 0 0 1 1 1 0 0 0 0 1 1 0 1 0 1 0 0 1 0 0 0 1 0
0 0 1 0 1 1 1 0 1 1 0 1 1 0 0 1 1 1 0 0 0 0 1 1 0 1 0 1 0 0 1 0
0 0 1 0 0 0 1 0 1 1 1 0 1 1 0 1 1 0 0 1 1 1 0 0 0 0 1 1 0 1 0 1
0 1 0 1 0 0 1 0 0 0 1 0 1 1 1 0 1 1 0 1 1 0 0 1 1 1 0 0 0 0 1 1
0 0 1 1 0 1 0 1 0 0 1 0 0 0 1 0 1 1 1 0 1 1 0 1 1 0 0 1 1 1 0 0
1 1 0 0 0 0 1 1 0 1 0 1 0 0 1 0 0 0 1 0 1 1 1 0 1 1 0 1 1 0 0 1
1 0 0 1 1 1 0 0 0 0 1 1 0 1 0 1 0 0 1 0 0 0 1 0 1 1 1 0 1 1 0 1
1 0 0 0 1 1 0 0 1 0 0 1 0 1 1 0 0 0 0 0 1 1 1 0 1 1 1 1 0 1 1
1 0 1 1 1 0 0 0 1 1 0 0 1 0 0 1 0 1 1 0 0 0 0 0 1 1 1 0 1 1 1
0 1 1 1 1 0 1 1 1 0 0 0 1 1 0 0 1 0 0 1 0 1 1 0 0 0 0 0 0 1 1 1
0 1 1 1 0 1 1 1 1 0 1 1 1 0 0 0 1 1 0 0 1 0 0 1 0 1 1 0 0 0 0 0
0 0 0 0 0 1 1 1 0 1 1 1 1 0 1 1 1 0 0 0 1 1 0 0 1 0 0 1 0 1 1 0
0 1 1 0 0 0 0 0 0 1 1 1 0 1 1 1 1 0 1 1 1 0 0 0 1 1 0 0 1 0 0 1
1 0 0 1 0 1 1 0 0 0 0 0 0 1 1 1 0 1 1 1 1 0 1 1 1 0 0 0 1 1 0 0
1 1 0 0 1 0 0 1 0 1 1 0 0 0 0 0 0 1 1 1 0 1 1 1 1 0 1 1 1 0 0 0
```

Table 3–4: Frequency bands and data rates

Band	Chip rate	Modulation	Bit rate	Symbol rate
868–868.6 MHz	300 kchip/s	BPSK	20 Kb/s	20 K
902–928 MHz	600 kchip/s	BPSK	40 Kb/s	40 K
2400–2483.5 MHz	2 Mchip/s	O-QPSK	250 Kb/s	62.5 K

The symbol rate is 62.5 Ksymbols/s, or alternatively, 250 Kb/s. Because each symbol is represented by a 32-chip sequence, the chip rate is 32 times the symbol rate, i.e., the chip rate is 2 Mchips/s. In offset QPSK, the Q-phase is delayed compared with the I-phase by one-half of a chip period with respect to the I-phase chip sequence. Half-sine pulse shaping is used, where

$$p(t) = \sin\frac{\pi t}{2T_c} \text{ when } 0 \le t \le 2T_c, \text{ and } p(t) = 0 \text{ otherwise.} \qquad \text{Eq. 3–4}$$

Figure 3-22 shows an example of chip sequences with half-sine pulse shaping. The synchronization header consists of the concatenation of 32 binary zeros and the 8-bit start-of-frame delimiter 11100101. The PHY header contains the number of bytes in the payload. The maximum packet size is 127 bytes. The preamble sequence is used by the transceiver to obtain chip and symbol synchronization with an incoming message; it also provides a settling period for automatic gain control (AGC) circuitry.

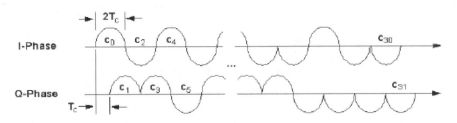

Figure 3–22: Example of chip sequences with half-sine pulse shaping

In the 868 MHz band, the data rate is 20 Kb/s, and in the 915 MHz band, 40 Kb/s. DSSS and BPSK are used in these bands. First, the bits are subject to differential encoding, which is the exclusive OR operation of a raw bit with the previous encoded bit. Then each bit is mapped to a 15-chip PN sequence. If the input bit is zero, it is mapped to 1 1 1 1 0 1 0 1 1 0 0 1 0 0 0. If the bit is one, it is mapped to 0 0 0 0 1 0 1 0 0 1 1 0 1 1 1.

The chip sequence is modulated using BPSK with raised cosine pulse shaping with a roll-off factor equal to 1. This pulse shape is described by the following equation:

$$p(t) = \frac{\sin(\pi t/T_c)}{(\pi t)/T_c} \frac{\cos(\pi t/T_c)}{1 - 4t^2/T_c^2} \qquad \text{Eq. 3-5}$$

Note that the 2 Mchip/s OQPSK waveform of the IEEE 802.15.4 2400 MHz air interface, with 6 dB bandwidth of about 1.5 MHz, and the 600 kchips/s BPSK waveform of the 902–928 MHz air interface, with bandwidth of about 600 kHz, both meet the definition of digital modulation.

The chip rate is 300 Kchips/s for the 868 MHz band and 600 Kchips/s for the 915 MHz band. This chip rate is 15 times higher than the symbol rate.

The PHY must be able to perform clear channel assessment (CCA). There are three ways to perform CCA—energy thresholding, IEEE 802.15.4 signal detection, and a combination of both (IEEE 802.15.4 signal detection with energy above a threshold).

IEEE 802.15.4 MAC

The main features of the MAC sublayer are beacon management, channel access mechanism, dynamic channel selection, frame reception and acknowledgment, association, disassociation, and security. Security functions are necessary for IEEE 802.15.4 to succeed in the market. Therefore, access control lists and symmetric cryptography are supported.

The LR-WPAN uses two types of channel access, depending on the network configuration. Networks without a beacon use a CSMA/CA channel access mechanism. This CSMA/CA channel access mechanism is much like the mechanism in IEEE 802.11.

Networks with a beacon use a slotted CSMA/CA channel access protocol. The slotted CSMA/CA mechanism operates as follows. A station that wants to transmit senses the channel. If the channel is idle it will wait for a random number of slots and begin transmitting on the next available slot boundary. Backoff slots are aligned with the start of the beacon. The beacon itself and acknowledgments are sent without CSMA/CA.

There are three types of data transfer. The first two are from a RFD to a FFD and from a FFD to a RFD. One is from the network device to the network coordinator. The second is from the network coordinator to the network device. The mechanisms for each transfer type depend on whether the network is beacon-enabled or not. When a network device needs to transfer data to the network coordinator in a beacon-enabled network, it first listens for the network beacon. The beacon is necessary so that the network device can determine when it can transmit. In a non-beacon-enabled network, the network device directly uses the CSMA protocol to transmit its data to the network coordinator. The network coordinator acknowledges a successful transmission. The acknowledgment policy can be no acknowledgment.

The network coordinator can send data to a network device in one of the following two ways. In a beacon-enabled network, the coordinator indicates pending data for a network device in its network beacon. The network device requests the data and then the network coordinator sends the appropriate data packet. The network device acknowledges successful transmission. In a non-beacon-enabled network, the coordinator must keep the data until the appropriate network device makes contact and requests the data.

A network device can contact the network coordinator in two ways. First, the network device can poll the network coordinator at a specific rate. If data is pending, the network coordinator transmits the data. The second way follows data transfer from the device to the coordinator. Then the network coordinator can indicate in its acknowledgment that it has data for the network device.

Peer-to-peer data transfers follow the CSMA scheme.

The packet structures have been designed to keep the complexity to a minimum while at the same time making them sufficiently robust for transmitting on a noisy channel. The LR-WPAN defines four frame structures: beacon frame, data frame, acknowledgment frame, and a MAC command frame. The beacon frame contains a beacon payload.

Every MAC frame consists of a MAC header, variable-length payload, and a footer consisting of the frame check sequence (FCS). The MAC header comprises frame control and addressing information. The frame check sequence contains an industry-standard 16-bit cyclic redundancy code, specified by the generator polynomial $G(x) = x^{16} + x^{12} + x^5 + 1$.

Suppose that $M(x) = b_0 x^{k-1} + b_1 x^{k-2} + ... + b_0$ is the polynomial representation of the sequence of bits that we want to perform CRC on. First, $M(x)x^{16}$ is divided by $G(x)$ modulo 2 to obtain $R(x)$. The coefficients of $R(x)$ are the value of the CRC. The MAC header contains the destination and source addresses and the CRC bit field.

Before devices can exchange data, a network must be formed. A low-rate WPAN can be started by a full-function device. The device that wants to start a PAN must first perform an active or passive channel scan. The active channel scan consists of transmitting the beacon request command in every channel. If there are network coordinators that are active, they, in response to a beacon request command, will start transmitting beacons. A passive channel scan is when a device simply listens for beacons in every channel. After completing the scan, if there are no coordinators in the channel where it wants to operate, the device must select a suitable PAN identifier and will start to consider itself as a network coordinator.

Device discovery is always important in PAN, because it enables a variety of applications. To associate with a network, a device does a scan to find existing networks to join. If more than one network is operating, the device must select which PAN to join. The algorithm according to which a PAN is selected is not specified and is left to the application developers. After selecting a network, the device can initiate association by sending an association request command during the contention access period of an existing PAN. When the coordinator receives the association request it will decide whether there are enough resources available for the device and assign a 16-bit address to the device.

Two types of access mechanisms are used, depending on whether the network is beacon-enabled or not. The beacon contains information about the structure of the following frame. The beacon is transmitted without the use of CSMA. In the beacon, the network coordinator device provides information about the permitted packet size, the current network coordinator operational state, and the length of the contention period. Because these parameters can change, the network coordinator must keep the information contained in the beacon up-to-date.

The contention period, in which the medium access mechanism is CSMA/CA, starts immediately after the beacon. In the contention period all frames, except acknowledgments, use CSMA/CA to access the channel. Considering the CSMA/CA mechanism for low-rate WPANs, the MAC sublayer keeps track of the number of retry attempts. Initially the number of retry attempts is set to 0. The MAC then issues a clear channel assessment request to the physical layer. The physical layer senses the channel to determine whether it is free or busy. If no activity is detected, the device may transmit. If activity is detected the device backs off for some random period before beginning the CSMA/CA procedure again. If the channel is idle then the MAC will check whether the backoff timer has expired before transmitting. At every time slot the backoff timer is decremented by one, and clear channel assessment is performed. If the channel is idle and the backoff timer has expired the device can transmit. The MAC must ensure that the entire transmission, including an acknowledgment, if one is requested, can be accomplished before the end of contention access period. If the channel is busy, the number of backoff attempts is incremented by one. If this value is greater than the maximum number of backoff attempts then the transmission is considered to have failed. Acknowledgements and retransmissions are optional. Every frame has a field, called the acknowledgment field. If this bit is set, acknowledgment is required. Otherwise it is not. If acknowledgment is not required the frame is never retransmitted. Otherwise retransmissions will be performed, up to a certain maximum.

The LR-WPAN standard allows the optional use of a superframe structure. The super frame format is defined by the network-coordinator. A superframe consists of beacons, contention access period, and contention-free period as shown in Figure 3–23.

While this is similar to the superframe used in 802.15.3, the contention-free period is optional. A superframe may have active and inactive portion, as illustrated in Figure 3–24. During the inactive portion of a superframe, devices can enter low-power mode. The beacons are sent by the network coordinator and are used to synchronize the devices and to identify the network. The channel access mechanism during the contention period is slotted CSMA/CA. All transactions must be completed by the time of the next network beacon. For low latency applications or applications requiring a specific data bandwidth, the network-coordinator may allocate dedicated slots, called guaranteed time slots (GTSs). The GTSs comprise the contention-free period. The number of guaranteed time slots in a frame is up to the network. If there is a contention-free period, all contention-based transactions must be completed before the first allocated slot begins. Also each device transmitting in an allocated slot must ensure that its transaction is complete before the next allocated slot or the next beacon begins. The beacon contains information about the length of the contention and contention-free periods. A beacon-enabled network is used for supporting low-latency devices such as PC peripherals. If the network does not need to support such devices, it can elect not to use the beacon for normal transfers. However, the beacon is still required for network connection purposes.

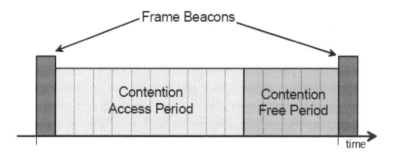

Figure 3-23: Superframe structure including an optional CFP

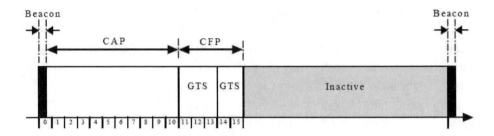

Figure 3-24: General superframe structure, with active and inactive periods

The GTSs follow immediately after the contention period. Within the GTS devices do not use CSMA/CA. GTSs are allocated exclusively for communication between the coordinator and a device. The device address and direction of communication (to or from the coordinator) uniquely identifies a GTS. GTS are allocated on a first-come, first-served principle. At any one time, there may be up to seven GTSs, because the device address field is three bits long. QoS can be changed on the fly—GTSs are allocated and deallocated dynamically at the discretion of the network coordinator. A device that has a GTS can also operate in the CAP. Only full-function devices can have GTSs, reduced-function devices cannot have GTSs. One full-function device can have one transmit GTS and/or one receive GTS. A device can request the allocation and deallocation of GTS. In response to a request, the network coordinator will grant a GTS if capacity is available. Note that after the deallocation of a GTS, the CFP can become fragmented. The network coordinator must ensure that there are no holes left by redundant slots. It may have to reallocate the GTS to make the contention-free period one contiguous block with no gaps. Reallocation is performed by appropriately informing the concerned devices. The set of slots must complete before the start of the next superframe. GTSs are not equal in length, and in this way devices with different bandwidth requirements can be accommodated.

In a network that does not use beacons, the entire period can be considered a contention access period. Devices operating in a non-beacon-enabled PAN will poll the coordinator for data on a periodic basis.

The access mechanism provides two types of QoS configurations. The first one is the priority in queuing. It is available in the contention-access period. The priority flag in the frame control field of the MAC header indicates the priority of the higher layer assigned to the packet. The priority types are high and normal. The packet marked high should transmit before the normal packets if both types of packets are in the queue. Beacon-enabled networks can also have a contention-free period, which provides the second QoS mechanism.

Several features of the protocol enable very low power consumption—lower than any other IEEE 802 standard. A superframe has an active part and an inactive part. Devices can enter low-power mode during the inactive part. The boundary between the active and inactive part can be adjusted.

Because the network is wireless, it is possible for a device to stop receiving beacons from the coordinator for an extended period of time. Devices that become lost should not transmit to avoid conflicts on the wireless medium. A lost device can do an orphan scan to realign itself. The lost device sends an orphan notification command on each channel. If a coordinator receives the orphan notification command, it will check its device list to determine whether the orphaned device is a member of its PAN. If the device is a member of its PAN, the coordinator will send a coordinator realignment command back to the orphaned device. If the orphan scan is unsuccessful, the device will assume that it has permanently lost the connection.

Security

The MAC is responsible for providing security services. The three security modes in the order of increased security are: the unsecured mode, ACL mode, and secured mode. The unsecured mode is mandatory to implement. ACL mode and secured mode are optional, but devices that support secured mode also must support ACL mode. Security can be delegated to the higher layers. Unsecured mode does not provide any security.

Access control is a method where a device can select which devices it is willing to communicate with. In ACL mode, devices check for every frame even if it is received from a device in the access control list. The access control list is composed of devices with which a key is shared. There are no cryptographic operations in ACL mode. In ACL mode, there is only the access control security service. The secured mode has all four security services: access control, data encryption, frame integrity, and sequential freshness. However, key management and device authentication are outside of the 802.15.4 standard and can be done only at higher layers.

Each device that implements security must support one or more security suites. The security suite indicates the symmetric cryptographic algorithm and integrity code.

All security suites use AES. Secured mode also uses the access control list, but in addition, includes cryptographic protection. In secured mode, the access control list is checked when a frame is transmitted. Frames are not sent to a device not in the access control list. Every received frame is also checked to see whether it comes from a device in the access control list. Furthermore, when a frame is sent, the security suite is checked from the access control list. If the security suite defines encryption, encryption is performed on the MAC payload.

If the security suite defines integrity code, the integrity code is applied to the MAC header concatenated with the MAC payload. The result of the integrity code computation is placed in the payload in addition to any other data in the payload. Finally, the FCS is computed over the modified frame. At the receiver, first, the FCS is checked. Then, if the security suite defines encryption, the decryption operation is performed on the payload. Data encryption protects data from being read by parties without the cryptographic key. Data encryption may be performed on beacon payloads, command payloads, and data payloads. If the security suite defines an integrity code, the integrity code is checked. If any of the security operations fail, an error is generated. Similar to IEEE 802.11 and IEEE 802.15.3, frame integrity is a security service that uses a message integrity code (MIC) to protect data from being modified by parties without the cryptographic key. Integrity can be provided on data frames, beacon frames, and command frames. Sequential freshness is a security service that uses an ordered sequence of inputs to protect data with an old sequence value from being replayed. This provides evidence that the received data is newer than the last data received.

Security suites supported by 802.15.4 are described next. A security suite is a set of operations. The name of security suite indicates the symmetric cryptographic algorithm, mode, and integrity code bit length. The length of the integrity code is less than the block size of the cryptographic algorithm and it indicates the probability that a random guess of the integrity code would be correct. There are

Chapter 3: Standards for Wireless Personal Area Networking (WPAN)

seven security suites defined in the standard: AES-CCM-64 is mandatory to implement. Six more are optional.

The first security suite is AES-CTR. In the AES-CTR symmetric encryption algorithm, a key stream is generated first using a key and a nonce. Each block in the key stream is different from all other blocks. A block cipher is applied to the set of input blocks to produce a sequence of output blocks. These output blocks are XORed with the plaintext to produce the ciphertext. This encryption is symmetric and the decryption operation consists of the generation of the key stream and the XOR of the key stream with the ciphertext to recover the plaintext. Because all operations are performed on blocks of data this algorithm lends itself very well for a parallel implementation. AES-CTR provides access control, encryption, and sequential freshness, but it alone cannot provide frame integrity.

Frame integrity can be provided by the CBC-MAC symmetric authentication algorithm. CBC-MAC can also provide access control, but it alone cannot provide sequential freshness and encryption. The CBC-MAC security suite consists of performing authentication on the encoding of the length of the combined header and payload, concatenated with the header, concatenated with the payload. The CBC-MAC used in 802.15.4 consists of the generation of an integrity code using a block cipher in CBC mode. The CBC-MAC algorithm makes use of an underlying block cipher to provide data integrity on input data. The block cipher transforms (or encrypts) input vectors of the block size. The data to be authenticated is grouped into contiguous blocks, each with length equal to the block size. The verification operation consists of the computation of the integrity code and comparison to the received integrity code. Depending on the size of the integrity code, 128-bit, 64-bit, and 32-bit CBC-MAC algorithms are supported by IEEE 802.15.4.

Like IEEE 802.11, IEEE 802.15.4 is using the combined symmetric encryption and authentication algorithm called CTR + CBC − MAC (CCM) [B144] provides all four security services. Depending on the integrity code, 128-bit, 64-bit, and 32-bit CCM is supported. The block size in all cases is 128 bits. Extensions to other block sizes are possible, but are not straightforward. CCM is a generic authenticate-and-

 Note that in the CTR encryption algorithm, cipher text is produced from plain text, and the message integrity check is not encrypted.

Wireless Communication Standards 215

encrypt block cipher mode. The CCM symmetric encryption and authentication consist of the generation of an integrity code followed by the encryption of plaintext data and the integrity code. The output is the encrypted data and the encrypted integrity code.

There are two parameters, M (the size of the authentication field) and L (the size of the length field). Larger values of M ensure greater protection against an attacker who can undetectably modify a message. At the same time, larger values of M lead to a larger message size. In 802.15.4 M is between 4 and 16 bytes and L is between 2 and 8 bytes. This accommodates a variety of applications. In CCM-128, L is 2 bytes and M is 16 bytes, in AES-CCM-64, M is 8 bytes and L is 2 bytes. In AES-CCM-32, M is 4 bytes, and L is 2.

To encrypt a message, the sender needs (1) an encryption key K and (2) a nonce of $N = 15 - L$ bytes. The nonce must be unique for the given encryption key. The message is divided into blocks of $l(m)$ bytes. There can be additional authenticated data that is not encrypted. The first step is to perform authentication. This is done using CBC-MAC. Then the message is encrypted with the CTR algorithm. The final result is the concatenation of the encrypted message, followed by the encrypted authentication value.

Coexistence issues involving IEEE 802.15.4

Clearly, IEEE 802.15.4 is yet another technology that operates in the 2400 MHz band. The channelization scheme of 802.15.4 in the 2400 MHz band is shown in Figure 3–25. As a result, a systematic investigation of coexistence is too complex. Because IEEE 802.15.2 does not consider IEEE 802.15.4, the coexistence performance of IEEE 802.15.4 with respect to IEEE 802.11b, IEEE 802.15.1 and IEEE 802.15.3 is analyzed in a separate annex to the standard. Generally IEEE 802.15.4 devices are expected to coexist well with other wireless systems.

The coexistence characteristics of 802.15.4 include the following:

- The medium access mechanism is CSMA-CA to avoid interference. In particular performing clear channel assessment by energy thresholding increases the coexistence performance. Note that in all coexistence scenarios combining energy level information and signal-to-noise ratio information provides information whether the packet errors are a result of low signal levels or strong interference.

- To make the protocol frequency agile, 802.15.4 specifies dynamic channel selection (DCS). 802.15.4 devices will work in that portion of the ISM frequency bands that is free of interference.
- 802.15.4 uses direct-sequence spread-spectrum, which is robust, since the spreading factor is high. This will reduce substantially the impact of other wireless systems to 802.15.4. In terms of interference to others, 802.15.4 will be perceived by other wireless systems as wideband noise.
- 802.15.4 devices will have very low duty cycles (less than 1 percent) and output power less than 0 dBm. Although government regulations in the United States and around the world allow output power up to 1 W in the 2400 MHz band, 802.15.4 devices will transmit at much lower power levels. This will eliminate the need for a filter to suppress the out-of-band emissions. The power levels will be between –3 dBm and 10 dBm, 0 dBm being typical.

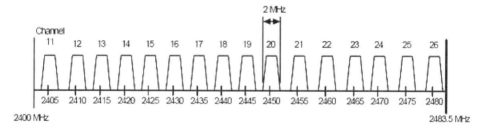

Figure 3–25: IEEE 802.15.4 channelization in the 2400 MHz band

The 1% Transmit Duty Cycle is the main limitation on operation within the 868.0–868.6 MHz band. However, it is the responsibility of the higher protocol layers to ensure that these parameters are satisfied.

In terms of impact to other wireless systems, the bandwidth of a IEEE 802.15.4 signal is wide and typically only a fraction of it will be received by other wireless devices. Because IEEE 802.15.1 uses frequency-hopping, the design of IEEE 802.15.4 ensures that interference will affect only three out of the 79 hopping frequencies. IEEE 802.11b will combat the interference from IEEE 802.15.4 by the processing gain of the IEEE 802.11b spreading process.

Dynamic channel selection is discussed next. In a star network, the central network controller initiates dynamic channel selection if the current QoS requirements cannot be satisfied due to adverse conditions of the channel. The controller collects channel status information from the network devices to arrive at this decision. The algorithm required to use the channel status information and decide change of channel is not specified. If the central network controller decides to change the channel, first it will look for other available channels. If it finds an available channel, it will begin sending beacons with the new channel information and the timeout for changing the channel. The devices that received a beacon with new channel information will change to the new channel within the indicated timeout duration and wait for beacons in the new channel.

For a cluster tree network, the cluster head gathers the information to determine if the channel quality is degrading. The cluster head keeps a list of available clear channels and makes the decision whether or not to initiate a channel change. As in a star network, the actual decision-making algorithm to determine whether or not a channel change is necessary is beyond the scope of this standard. If the cluster head decides to change the channel, the cluster head may transmit a message on the original channel that contains the new channel information to all members of the cluster. Because the original channel is impaired, new channel information is transmitted three times by each receiving device to increase the probability of reception. It is not expected that a receiving device will receive all three transmissions. After the three transmissions, devices change to the new channel.

In both star and cluster tree topologies, there may be devices that were unable to receive the new channel information and that will become lost after the channel switch.

WPAN TECHNOLOGY AND BUSINESS TRENDS

The first WPAN technology, Bluetooth, was developed because some cell phone manufacturers wanted to increase the value of their main product by making it capable of communicating with devices other than phones. At present high-end cell phones on the market have Bluetooth functionality. Within a couple of years almost all phones will have Bluetooth. The original intention of the Bluetooth SIG was for Bluetooth to achieve a price of $5 and become a truly ubiquitous technology. According to U.S. antitrust laws, price discussions within IEEE 802 cannot take place, and therefore, the 802.15 Working Group has never discussed prices. While the price target of $5 seems too aggressive, perhaps as a consequence of the fundamental principles of communications, an increasing number of products are expected to have Bluetooth. In many of these products Bluetooth will be embedded, which will be a feature differentiating them in the marketplace. Higher-priced devices can absorb the cost of Bluetooth, and as a result some say that the cost of Bluetooth is zero. It seems reasonable that Bluetooth will succeed as a cable replacement technology. Bluetooth PC cards and various computer peripheral devices are among the first to include Bluetooth, in addition to cell phones. Similar to 802.11, some laptops are manufactured with an integrated Bluetooth unit. PDAs such as Palm and Handspring increasingly come with wireless communication capability. It must be noted that some powerful PDAs come with 802.11. Outside of computing and communication devices, in the near future a number of other devices and equipment will have a number of Bluetooth units.

Recently 802.15.3 began work on a new physical layer 802.15.3a, providing data rates in excess of 100 Mb/s, and up to 480 Mb/s. This physical layer will be based on ultra-wide band (UWB) technology and will attempt to enable true "cable replacement". Ultra-wide band is one of the most over-hyped wireless technologies in the last few years and the industry interest in 802.15.3a is very large. The standardization work within 802.15.3a involves most of the major semiconductor companies such as Intel, Texas Instruments, and Motorola. Start-up companies focused on UWB technology such as Time Domain®, XtremeSpectrum®, and Staccato Communications, Inc.® also participate.

The concept of UWB originated about 100 years ago, due to Marconi. The modern UWB technology was developed in the 1980s and 1990s. The main motivation initially was military communication systems, since FCC rules did not allow UWB operation for commercial use outside of the military. Since UWB operates over a very wide bandwidth there was a concern about interference to licensed users "and other important radio operations". In February 2002, the FCC amended the Part 15 regulations, allowing UWB to operate over a band that is 7500 MHz wide (3100–10600 MHz) on an unlicensed basis with average radiated emission limit of –41.3 dBm/MHz. The peak emission limit is 60 dB above the average. Note that UWB is defined by the FCC as any signal having a fractional bandwidth greater than 20 percent and occupying a band of 500 MHz or more. Fractional bandwidth is the ratio of baseband bandwidth to RF carrier frequency. The band is bounded by the points that are 10 dB below the highest radiated emission. Note in particular that the modulation used is not relevant to the definition of UWB. The rules allow any modulation scheme to be used, as long as the fractional bandwidth is greater than 20 percent and the occupied bandwidth is greater than 500 MHz. At present other regulatory bodies around the world have not sanctioned the operation of UWB devices. While favorable rulings are expected, the allowed frequency band of operation may be different. (See Table 3–5 below.)

Table 3–5: FCC regulations for the mean EIRP of UWB systems

Frequency band (MHz)	Mean EIRP (dBm/MHz) (indoor/handheld)
960–1610	–75.3/–75.3
1610–1900	–53.3/–63.3
1900–3100	–51.3/–61.3
3100–10600	–41.3/–41.3
>10600	–51.3/–61.3

Traditionally UWB operation has been achieved with pulse-position modulation (PPM), where very short pulses are used. These pulses must have zero-mean, because an antenna cannot radiate a signal at zero frequency. If the pulse width is shorter compared to the delay spread then good performance in multipath channels can be achieved. In UWB, the coherence bandwidth is smaller than the signal bandwidth, and the coherence time is larger than the symbol duration [B117], [B139], [B140].

Initially there were 31 proposals submitted to 802.15.3a from various companies. The number of proposals was quickly reduced to two main technologies, as a result of dropouts and mergers. The first main proposal was based on impulse radio technology [B105]. This proposal uses BPSK and QPSK modulation with DSSS with 24 chips per symbol, leading to high spread-spectrum processing gain. It uses the entire band allowed by the FCC with the exception of the UNII frequencies in the 5GHz region. Although not mandated by the FCC, this is chosen to enable coexistence with WLAN devices. To satisfy the FCC regulations the baseband signal according to the proposal can be 1.368 GHz or 2.736 GHz. The second main proposal was a merger of several proposals. According to this proposal the baseband signal is 528 MHz. The entire 7.5 GHz-wide band is divided into four smaller bands, with frequency hopping between these smaller bands used to facilitate multiple access. The carriers are modulated using QPSK, followed by 128-point FFT. Like the first proposal, the second proposal also avoids using the 5 GHz UNII bands. The performance comparisons between these two main proposals follow many of the comparisons between single-carrier and multi-carrier technologies.

IEEE Std 802.15.3 and IEEE Std 802.15.4 have not yet made it to the mass-market stage and represent interesting opportunities. IEEE 802.15.3 together with 802.11 will be a competing technology to be used in consumer electronic products for the home, such as TV sets and digital cameras. Digital cameras are typically connected via a cable to a computer for image and video communication. This cable can be conveniently replaced by a wireless technology, and Bluetooth is not fast enough.

IEEE 802.15.4 seems a strong competitor in a new and fast-growing market—wireless sensors. Wireless sensors have many applications. 802.15.4 products can be implemented in a way that requires two AA batteries for one year of operation. Examples of embedded devices suitable for low-rate WPAN are PIR sensors, cordless switches, intelligent remote controls (network-device) and security, lighting and heating, ventilating, and air-conditioning (HVAC) controllers (network-coordinator). Similar sensors can be used on buildings, bridges, and other structures to monitor their seismic movements. In the near future wireless sensors will find a variety of applications in industrial control and monitoring, intelligent agriculture, supply chain management and asset tracking, health monitoring, security and military sensing. Some companies believe that low data-rate applications actually represent a bigger market opportunity than high data-rate wireless personal area networks. Some early technological leadership has been established by companies like Motorola, and several start-up companies including Staccato Communications of San Diego, CA, and S5, Inc., of Draper, UT. To promote this technology and ensure interoperability among products from different vendors, interested companies have formed an industry consortium called ZigBee Alliance™ of San Ramon, CA [B63]. The relationship between 802.15.4 and ZigBee is similar to the relationship between 802.11 and the Wi-Fi Alliance of Mountain View, CA.

Ultra-wide band technology is not only suitable for high data-rate applications. It is also suitable for low data-rate, low power applications. 802.15.4 is currently exploring another physical layer to be called 802.15.4a, which will likely be based on UWB technology [B3].

The WPAN devices can be implemented in two types of devices —embedded (non-host enabled) and host enabled. The embedded node architecture consists of a user interface, the actual device application, a network layer, and an on-air protocol stack. The API connects the device application with the network layer, providing a consistent application interface to all nodes. The network layer definition, the device application, and its user interface are outside the scope of the standard and will typically be provided by the end-application developer. The on-air protocol stack consists of a LLC sublayer, a MAC sublayer, and a PHY layer. A network-device may not need any of the upper layers.

For host-enabled devices to interface between the WPAN device and the external host a host stack is necessary. The interface can be via any suitable interconnect protocols such as RS232 or USB. Examples of host-enabled nodes are USB hubs and Internet portals. This architecture consists of a local user interface, the actual device application, an on-air protocol stack and a host stack providing inter-connectivity with the host. The interface between the application and the on-air protocol stack is via the API. The on-air protocol stack allows an application to access a WPAN. The user interface is dependent on the function of the application and the amount of information that needs to be communicated or extracted from the consumer. In network-devices, the user interface can be as simple as a single button, which can be used as a stimulus for certain networking functions, such as joining or leaving a network. Host enabled network-coordinator may require a local user interface. This will be used when the external host is not connected since the network-coordinator must still be able to function in the absence of a host. The user interface is not specified in this standard.

It is worth examining again the question of whether WPANs and WLANs are competing or complementing technologies. It would seem that WPAN devices used as cable replacement avoid competition with WLAN devices because of low power consumption and cost. However, WPANs can cover a much larger range if they transmit at the maximum allowed output power. Therefore, WPANs can be used for other applications, such as infrastructure computing and home networking. In these applications WPANs can potentially compete with WLAN technologies. However, this competition will likely be won by WLAN. 802.11 has established itself as the most successful wireless standard. Many consumers use 802.11 in their offices and when they bring their laptops home expect to be able to connect to other 802.11 devices. Therefore, 802.11 has an advantage for home networking applications as well. There are three 802.11 physical layers in the marketplace—802.11b, 802.11a, and 802.11g. The cost of 802.11b is falling rapidly, and is approaching the cost of WPAN devices. The cost of 802.11a and 802.11g is also expected to decrease in the next couple of years. In the near future consumer electronics devices such as TV sets, DVD players, MP3 players, musical instruments, etc., will have the capability for wireless communication. In the long run, other consumer devices and kitchen appliances will have wireless communication capability. Since on one hand the majority of these devices are mains-powered, while on the other hand the range is relatively small, it remains to be seen whether vendors will choose WPAN over WLAN.

Chapter 4 Air Interface for Fixed Broadband Wireless Access Systems

OVERVIEW

Over the last several years, the area of broadband data communications experienced very high growth. The term "broadband" is usually taken to mean the capability to deliver significant bandwidth to each user, much higher than narrowband voiceband modems. Following ITU terminology, the term "broadband" means transmission rates greater than 1.5 Mb/s. Broadband Internet access became a mass-market industry reaching most of the 100 million homes and 3 million businesses in the United States. Broadband access reaches an even greater percentage of homes and businesses in some other industrialized countries, such as Korea and Canada. For a relatively long time now, it has been known that the major technologies for broadband Internet access are DSL, cable, wireless, and optical fiber communications. Of these four technologies, two—DSL and cable—are mature technologies and have reached the mass-market stage. Optical communications provides very high data rates, but the cost of laying optical fibers is extremely high, in fact, prohibitively high in urban areas where most of the customers are. Because of cost issues, it is not clear whether optical communications will ever become a mass-market technology. However, it is expected that broadband wireless access (BWA) systems will become a technology that is generally competitive with DSL and cable, connecting businesses and users' premises with core networks.

Clearly, if BWA systems are to compete with DSL and cable, their market growth will be significant for several reasons. First, in the long run, BWA systems are capable of delivering significantly higher data rates than DSL or cable. DSL technology requires customers to be close to the telephone company's central office. DSL can deliver up to 6 Mb/s at distances up to 18,000 feet. At present, most DSL subscribers get much lower speeds. Customers that are more than 18,000 feet away from the telephone company's central office cannot get DSL service. Similarly, cable broadband service delivers 1.5 Mb/s, but requires a TV cable. Though most of businesses and residential users in North America have access to either DSL or cable, some do not. Furthermore, many more business and residential users around the world cannot get DSL or cable. These users represent a significant potential market for BWA. In addition, compared with DSL and cable networks, the use of wireless techniques results in a number of benefits for users and for service providers. For service providers, these benefits include relatively low up-front costs. For consumers, the benefits include the convenience of a wireless connection.

What is the architecture of BWA networks? Fixed broadband wireless access systems typically include at least one base station (BS) and a number of subscriber stations (SS). In addition there may be links between base stations, repeaters, and possibly other equipment. Base stations provide connections to core networks on one side and radio connection to other stations on the other. A reference fixed BWA system diagram is provided in Figure 4–1.

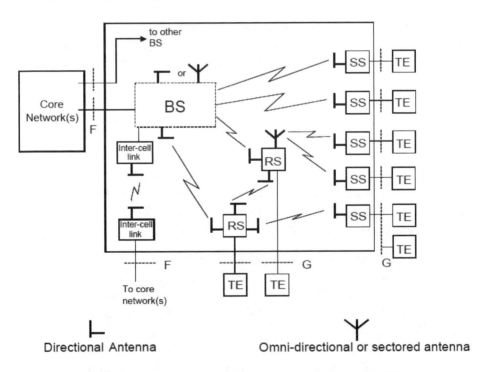

Figure 4–1: Fixed broadband wireless access (BWA)

Fixed BWA systems have multipoint architectures. The term multipoint includes point-to-multipoint (PMP) and multipoint-to-multipoint (MP-MP). PMP systems comprise base stations, subscriber stations and, in some cases, repeaters. On the uplink, the sending entity is the SS and the receiving entity is the BS. On the downlink, the sending entity is the BS and the receiving entity is the SS. Note that one difference compared with WLAN and WPAN is that antennas with a variety of radiation patterns are used. Base stations use relatively wide-beam antennas, divided into one or several sectors providing up to 360-degrees coverage with one or more antennas. The sectorized antenna of the base station is capable of handling multiple independent sectors simultaneously. Within a given frequency channel and antenna sector, all stations receive the same transmission. To achieve complete coverage of an area, more than one base station may be required. The connection between base stations is not part of the fixed BWA network itself. The connection is achieved by use of radio links, fiber optic cable, or equivalent means. Links between BSs may sometimes use part of the same frequency allocation as the fixed BWA itself.

Routing to the appropriate BS is a function of the core network. In general, subscriber stations use highly directional antennas that face the BS. Subscriber stations share use of the radio channel. This may be achieved by various access methods, including frequency division, time division, or code division. Some systems deploy repeaters. In a PMP system, repeaters are generally used to improve coverage to locations where the BSs have no line of sight within their normal coverage areas or, alternatively, to extend coverage of a particular BS beyond its normal transmission range. A repeater relays information from a BS to one or a group of SSs. A repeater may operate on the same downlink frequencies as those frequencies that it uses, facing the BS, or it may use different frequencies (i.e., demodulate and remodulate the traffic on different channels). MP-MP or mesh systems have the same functionality as PMP systems. In MP-MP systems, traffic may pass through one or more repeaters to reach a subscriber. Most stations are repeaters that also provide connections for local subscribers. Antennas are generally narrow-beam directional types with means for remote alignment.

These networks operate transparently, so users are not aware that services are delivered by radio. A BWA network provides connection to many user premises within a radio coverage area. It provides a pool of bandwidth, shared automatically among the users. Demand from different users is often statistically of low correlation, allowing the network to deliver significant bandwidth-on-demand to many users with a high level of spectrum efficiency. Significant frequency reuse is employed. The range of applications is very wide and evolving quickly. It includes voice, data, and entertainment services of many kinds. Each subscriber may require a different mix of services; this mix is likely to change rapidly as connections are established and terminated. Traffic flow may be unidirectional, asymmetrical, or symmetrical, again changing with time. In some territories, systems delivering these services are referred to as Multimedia Wireless Systems (MWS) in order to reflect the convergence between traditional telecommunications services and entertainment services.

In response to these market needs, the IEEE 802.16 working group on Broadband Wireless Access was established in 1999. The industry interest was significant and the first standard (IEEE Std 802.16-2001) was approved in December 2001 and published in April 2002 [B82].

The IEEE 802.16 standard specifies the MAC and PHY of fixed broadband wireless access systems. Similar to IEEE 802.11, the MAC layer is structured to support multiple physical layer specifications to give the standard flexibility. IEEE 802.16 supports point-to-multipoint and multipoint-to-multipoint communications in the 2–66 GHz band. In addition to technological and business reasons, the existence of multiple physical layers reflects the fact that the electromagnetic propagation between 2 GHz and 66 GHz is not uniform. Each physical layer is appropriate to a particular frequency range and application. IEEE 802.16 is intended to operate in three different frequency bands.

The first group of frequency bands of interest is the licensed bands between 10 GHz and 66 GHz. In this frequency range, the wavelength is very short, and the attenuation of electromagnetic waves by various terrain and human-generated structures is severe. Therefore, line-of-sight between transmit and receive antennas is required. Multipath is not an issue, because only the main path from the transmitter to the receiver will have significant energy. While multipath is not a limiting factor, thermal noise and/or interference are the main limiting factors to the performance of wireless systems in this frequency band. These systems are used only in an outdoor setting, and rain will increase the attenuation experienced by the electromagnetic waves. Therefore, a sizable link margin must be reserved for rain loss. At the same time, the desire for high data rates means that high-order modulation schemes must be used. In turn, these high-order modulation schemes require large signal-to-noise ratio (SNR) for satisfactory operation.

Another difference between IEEE 802.16 and IEEE 802.11 or IEEE 802.15 is that the range of interference typically exceeds the operating range. This is compounded by the fact that the rain cells producing the most severe rain losses are not uniformly distributed over the operational area. This creates the potential for scenarios in which the desired signal is severely attenuated but the interfering signal is not. While rain attenuation is a technical problem, operating in the 10–66 GHz band leads to two technical advantages. First, for the reasons outlined previously, multipath is negligible. Second, the frequency bands of operation in this physical environment are typically large, on the order of 25–28 MHz. This allows data rates in excess of 120 Mb/s.

The second group of frequency bands of interest is the licensed bands between 2 GHz and 11 GHz. In these bands, line of sight is not necessary. Multipath can be significant and appropriate measures must be taken. Because there may or may not be line of sight, the signal power can vary significantly. Retransmissions may be necessary because of the lossy behavior of the wireless medium. As a solution to these problems, the 802.16 standard provides advanced power management techniques, and the physical layer for operation between 2 GHz and 11 GHz can be based on OFDM.

The third group of frequency bands of interest is the unlicensed bands between 2 GHz and 11 GHz. While the physical characteristics of these unlicensed bands are the same as the licensed bands between 2 GHz and 11 GHz, there are two differences. Because these bands are unlicensed, first, there may be other users, which causes an interference problem. Second, regulations limit the output power. These problems require dynamic frequency selection and power management.

Figure 4–2 provides a block diagram of IEEE 802.16. The architecture follows general IEEE 802 guidelines. The MAC layer comprises three sublayers—convergence sublayer, common part sublayer, and security sublayer. The MAC includes a convergence sublayer to better handle the higher-layer protocols placed above the MAC. The central part of the MAC is the common part sublayer. Like in the other 802 standards, it handles channel access, connection establishment and maintenance, and QoS. The third sublayer is the security sublayer, providing authentication, secure key exchange, and encryption.

Chapter 4: Air Interface for Fixed Broadband Wireless Access Systems

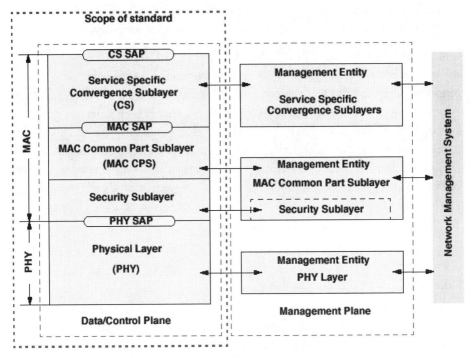

Figure 4–2: IEEE 802.16 Protocol structure

Although the MAC and PHY layers are independent, the integration between the MAC and the PHY is important for the overall system performance. For the MAC/PHY integration at a system level, two issues are important: the duplexing technique and whether the PHY is framed or nonframed. These choices should be made together with the PHY design. For a framed PHY, the MAC aligns its scheduling intervals with the underlying PHY framing. Better performance can be achieved with an unframed PHY, although it is not supported by 802.16. In this case, the MAC can choose the scheduling intervals to optimize system performance.

The upstream physical layer is based on the use of a combination of time division multiple access (TDMA) and demand assigned multiple access (DAMA). In particular, the upstream channel is divided into a number of "time slots." The number of slots assigned for various uses (registration, contention, guard, or user traffic) is controlled by the MAC layer in the base station and can vary in time for optimal performance. The downstream channel can be either based upon continuous time division multiplexing (TDM), where the information for each subscriber station is multiplexed onto the same stream of data and is received by all subscriber stations located within the same sector or in an alternative method (defined for the burst mode of operation), which allows bursts to be transmitted to specific SSs in a similar fashion to the TDMA upstream bursts.

The next three sections of this chapter are devoted to the MAC layer: the convergence, common part, and security sublayers. The fourth section beginning on page 256 is devoted to the necessary MAC enhancements for when the physical layer operates in the 2 GHz–11 GHz frequency range. Then the physical layers of IEEE 802.16 are discussed in the fifth section on page 269. Coexistence is very important for wireless access systems and is considered in the sixth section on page 317. The chapter concludes with an overview of the technology and business trends for BWA systems.

MAC CONVERGENCE SUBLAYER

Above IEEE 802.16, the higher layers will predominantly be ATM [B5] or IEEE Std 802.3 (Ethernet) [B91]. Therefore, two convergence sublayer specifications are developed—ATM and packet. Because the convergence sublayer is a separate part of the MAC, a vendor that wants to use a higher layer different than ATM or Ethernet can develop a different convergence sublayer. The convergence sublayer is not a physical, but a logical interface. In transmit mode, the convergence sublayer receives data from a higher layer, and classifies the data. Based on the classification, it can perform additional processing before delivering the data to the MAC common part sublayer. This type of additional processing can be payload header suppression (PHS). In receive mode, the convergence sublayer accepts data from the MAC common part sublayer. If the peer convergence sublayer has modified the data frame, the receiver will restore the frame before passing it on to the higher layer.

The ATM convergence sublayer (CS) is designed to support the different ATM services. It accepts ATM cells, classifies these cells, and maps them to MAC frames. It may also perform Payload Header Suppression (PHS). An ATM connection can be either Virtual Path (VP) switched or Virtual Channel (VC) switched. The ATM convergence sublayer differentiates between these two types of connections and assigns appropriate channel ID (CID) before delivering the data to the MAC common part sublayer. In this way, the ATM convergence sublayer guarantees the correct handling of the traffic by the MAC sublayer. The ATM header is thrown away, with the exception of the field indicating priority. Payload header suppression is the process of suppressing the repetitive portion of payload headers at the sender and restoring the headers at the receiver. It can be viewed as a compression method used to save bandwidth. A PHS Rule provides details on how a repetitive portion of the payload headers can be replaced with an appropriate index for transmission and subsequently regenerated at the receiving end. When payload header suppression is performed, a suppression mask can be used to allow select bytes not to be suppressed. This is used for sending bytes that change (such as IP sequence numbers) and still suppressing bytes that do not change. Payload header suppression is intended for unicast communications and is not defined in the multicast case. In addition to payload header suppression, multiple ATM cells, which share the same connection ID, may be packed together to further save bandwidth.

ATM interfaces are also characterized by the signaling mechanism that they use [B6]. The convergence sublayer does not support associated signaling, but supports common channel signaling (CCS). According to the common channel signaling mechanism for ATM networks [B6], signaling messages are carried over a connection completely independent of user connections. In addition, one ATM signaling channel can carry signaling messages for a number of user connections. Every IEEE 802.16 station must have a channel ID, corresponding to CCS messages. The questions regarding the mapping of ATM signaling messages are left to technology vendors to implement. This is one of the many issues that IEEE 802.16 leaves to vendors. In particular, designers of the base station MAC must map ATM signaling messages to corresponding MAC messages.

The second type of convergence sublayer is the packet convergence sublayer. It is used for the transport for all packet-based protocols such as Internet protocol (IP), point-to-point protocol (PPP) [B14], and IEEE 802.3/Ethernet. It performs similar function as the ATM convergence sublayer, including suppression of payload header information and rebuilding of any suppressed payload header information.

Both the ATM and the packet convergence sublayers perform classification of the higher-lay data. This classification is the process by which a MAC data is associated with a particular connection. This process facilitates the delivery of MAC data with the appropriate QoS constraints. In general, a classifier is a set of matching criteria applied to each packet entering the IEEE 802.16 network. There are multiple classifiers, applied according to a certain priority. If more than one classifier matches the packet, the highest priority classifier is chosen. The destination MAC address, the source MAC address, and other parameters can be used for the classification. When Internet protocol is used, the IP headers can be included in the classification parameters. Each classifier also defines whether or not payload header suppression is performed. The output of the classification procedure is the connection identifier, or connection ID. The concept of a connection, or service flow, is central in IEEE 802.16. The connection ID can be considered a connection identifier even for nominally connectionless traffic like IP, because it serves as a pointer to destination and context information. This connection identifier is used to distinguish between multiple uplink channels, all of which are associated with the same downlink channel. In IEEE 802.16, the connection ID is 16 bits long. The use of a 16-bit connection ID permits a total of 64K connections within each downlink and uplink channel. It is possible that a packet does not match any of the classifiers. In this case, the output of the classification procedure is not standardized and is left at the discretion of the vendors. The packet can be associated with a default connection ID or discarded.

The operation of the convergence sublayer can be summarized as follows. After the convergence sublayer of the sending entity receives a packet from the higher layer, it applies its list of classifier rules to determine the service flow, connection ID, and whether to turn on or off payload header suppression. If suppression is enabled, the sending entity suppresses all the bytes except the bytes masked by the mask. The convergence sublayer then prefixes the frame with the payload header suppression index and sends the block of data to the MAC SAP for further processing. When the packet is received, the receiving entity will determine the associated connection ID by examination of the generic MAC header. The receiving convergence sublayer then reassembles the packet.

MAC COMMON PART SUBLAYER

The common part sublayer is the central part of the IEEE 802.16 MAC. It defines the multiple-access mechanism.

> **Messages can be individually addressed, multicast, or broadcast.** Multicast messages are addressed to a group of subscriber stations, and broadcast messages are addressed to all stations.

In the downlink, the base station is the only transmitter that is operating. Therefore, it does not have to coordinate its transmissions with other stations. In the downlink, the base station broadcasts to all stations. Stations check the address in the received messages and retain only those addressed to them. In the uplink direction, the user stations share the channel. To determine which subscriber station has the right to transmit, three basic principles are used—unsolicited bandwidth grants, polling, and contention procedures. The method for sharing the channel is dynamic and functions on an on-demand basis. Sharing is on demand to give the service providers opportunities to differentiate themselves in the marketplace. Depending on the class of service used, the SS may be issued continuing rights to transmit, or the right to transmit may be granted by the base station after receipt of a request from the user.

The duplexing techniques supported by IEEE 802.16 will be considered first. Duplexing here means how the downlink is separated from the uplink. Compared with the other wireless standards, IEEE 802.16 follows a much more centralized architecture. This is appropriate for a technology that does not have to support ad-hoc networks. There are two ways to separate downlink from uplink—in the frequency domain and in the time domain.

In frequency division duplexing (FDD), downlink and uplink transmissions are performed at different frequencies and therefore may overlap in time. Although theoretically the width of the frequency bands devoted to downlink and uplink transmissions can be adjusted, this is very difficult to accomplish in practice. As a result, in FDD the asymmetry between uplink and downlink is static. A fixed duration frame is used for both uplink and downlink transmissions. This facilitates the use of different modulation types. It also allows simultaneous use of both full duplex subscriber stations, which can transmit and receive simultaneously, and half duplex subscriber stations, which cannot. A half duplex subscriber station can listen to the downlink channel only when it is not transmitting in the uplink channel. If half duplex subscriber stations are used, the bandwidth controller does not allocate uplink bandwidth for a half duplex subscriber station at the same time that it is expected to receive data on the downlink channel. The fact that the uplink and downlink channels use a fixed duration frame simplifies the bandwidth allocation algorithms. Figure 4–3 illustrates the bandwidth allocation in FDD.

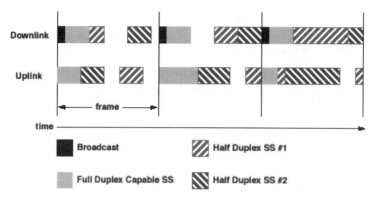

Figure 4–3:   Example of burst FDD bandwidth allocation

Time-division duplexing (TDD) divides time into uplink and downlink transmission periods, and therefore downlink and uplink may be performed at the same frequency band. In the case of TDD, a frame also has a fixed duration and contains one downlink and one uplink subframe. The frame is divided into an integer number of physical slots, which help to partition the bandwidth easily. Because in TDD, the balance between the amount of time devoted to uplink and downlink can be dynamically adjusted, the asymmetry between downlink and uplink is dynamic. The split between uplink and downlink is a system parameter and is controlled at higher layers within the system. Figure 4–4 illustrates the TDD frame structure.

A MAC data frame in IEEE 802.16 consists of three parts—header, payload, and cyclic redundancy code (CRC). The payload and the CRC are optional. The payload information may vary in length. Two MAC header formats are defined—generic and bandwidth request. The generic header is used if the payload contains MAC management messages, or data from the higher layers that has passed through the convergence sublayer. Bandwidth request frames do not have payload and are identified by the bandwidth request header. A particular connection ID may require a CRC to be attached. If present, the CRC covers the generic MAC header and the MAC payload. The CRC is calculated after encryption; i.e., the CRC protects the generic header and the ciphered payload.

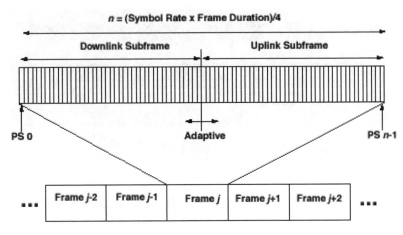

Figure 4–4:   TDD frame structure

Multiple MAC frames may simply be concatenated into a single transmission in either the uplink or downlink directions. MAC management messages, user data, and bandwidth request frames may be concatenated into the same transmission.

Fragmentation and packing are dual processes. Fragmentation is the process by which one MAC frame is divided into more frames. This process allows efficient use of available bandwidth relative to the QoS requirements. Every frame contains a sequence number, which allows the receiver to reconstruct the original frames and to detect the loss of any intermediate packets.

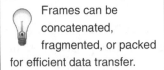
Frames can be concatenated, fragmented, or packed for efficient data transfer.

Packing is the opposite of fragmentation. If packing is turned on for a connection, the MAC may pack multiple frames into a single MAC frame. If ARQ is not being used, packing fixed-length frames is straightforward, because the receiving side can unpack simply by checking the length field in the MAC header and the entire length of the data. However, when packing variable-length MAC frames, where one MAC frame ends and another begins must be indicated. To accomplish this, the MAC attaches a packing subheader to each MAC frame.

The type field in the MAC header indicates the presence of packing subheaders. Packing frames on ARQ-enabled connections is very similar. The packing subheaders will be used by the ARQ protocol to identify and retransmit lost fragments.

Note that packing and concatenation are very similar.

The ultimate goal of fragmentation and packing is the efficient use of the air link. Fragmentation and packing also allow the MAC to tunnel various higher-layer traffic types without knowledge of the formats or bit patterns of those messages.

MAC management messages are used to describe the downlink and the uplink, and to handle ranging, registration, and privacy.

To describe the downlink and uplink, the BS transmits downlink channel descriptor (DCD) and uplink channel descriptor (UCD) messages at periodic intervals to define the physical layer characteristics of the downlink and uplink channels, correspondingly. The DCD and UCD messages contain the burst profile, which has information about the modulation type, forward error-correction type, preamble length, etc. The downlink map (DL-MAP) and the uplink map (UL-MAP) messages define the burst start times on the downlink and the uplink, respectively.

Another management function is ranging. Subscriber stations transmit ranging requests at initialization and periodically at the request of the base station to determine power and burst profile changes.

To efficiently support these management functions, at SS initialization, three management connections in each direction (uplink and downlink) are established between the SSs and the BS. These connection identifiers reflect the fact that there are inherently thee different QoS of management traffic between SSs and the BS. The basic connection is established during initial subscriber station registration and is used to transport MAC management messages with short delay. The primary management connection is for the exchange of longer, more delay-tolerant MAC management messages. Finally, the secondary management connection is used for delay tolerant, standards-based (DHCP, TFTP, SNMP, etc.) management messages.

Network entry and initialization

The procedure for initialization of a subscriber station can be divided into the following phases:

1) *Scan for downlink channel and establish synchronization with the BS.* Typically, subscriber station devices will have nonvolatile storage in which the last operational parameters are stored. In this case, these devices will first try to reacquire the downlink channel used previously. If this fails, the SS will begin to continuously scan the possible channels of the downlink frequency band of operation until it finds a valid downlink signal.

2) *Obtain downlink and uplink parameters.* The BS generates uplink channel descriptor (UCD) and downlink channel descriptor (DCD) messages on the downlink at periodic intervals. The BS generates UL-MAP and DL-MAP at intervals as specified in a particular PHY specification. These messages are addressed to all subscriber stations. First, to obtain downlink parameters, the MAC sublayer will search for the DL-MAP MAC management messages. The SS achieves MAC synchronization once it has received at least one DL-MAP message. An SS MAC remains in synchronization as long as it continues to successfully receive the DL-MAP and DCD messages. To obtain uplink parameters, the SS will wait for a UCD message from the BS in order to retrieve a set of transmission parameters for a possible uplink channel. These messages are transmitted periodically from the BS for all available uplink channels and are addressed to the MAC broadcast address. The SS can determine whether it may use the uplink channel from the channel description parameters. The SS collects all UCDs, which are different in their channel ID field, to build a set of usable channel IDs. The subscriber station will determine from the channel description parameters whether it may use the uplink channel. If the channel is not suitable, then the SS will try the next channel ID until it finds a usable channel. If the channel is suitable, the SS will extract the parameters for this uplink from the UCD. It will then wait for the next DL-MAP message and extract the time synchronization from this message. Before it can begin transmitting, the SS will wait for a bandwidth allocation map for the selected channel. Transmissions from the SS in the uplink will be in accordance with its MAC operation and the bandwidth allocation mechanism. The SS MAC is considered to have valid uplink parameters as long as it continues to successfully receive the UL-MAP and

UCD messages. If at least one of these messages is not received within certain time intervals, the SS will not use the uplink.

3) *Perform ranging.* Ranging is the process of acquiring the correct timing offset to align the SS's transmissions to a symbol marking the beginning of a mini-slot boundary. After the SS has received the UL-MAP message, it will scan the message to find an initial maintenance interval. The initial maintenance interval specified by the BS is large enough to account for the variation in delays between any two SSs. The maximum delay that one SS can experience is the maximum round-trip propagation delay due to cell radius plus the maximum allowable implementation delay. As a result, there could be significant variation in delay among subscriber stations. The ranging request message from the SS will be sent at the first initial maintenance transmission opportunity. The connection ID is set to zero in this case, because the SS does not have any connection IDs assigned to it. The first ranging request will be sent at minimum power level. This is an important step to minimize potential interference when a connection is being set up. If the first ranging request attempt is not successful, then at the next initial maintenance transmission opportunity, the SS will resend it at a power level, which is one step higher. This process will continue until the transmission is successfully received. Once the BS successfully receives the ranging request message, it will return a ranging response message addressed to the individual SS. The ranging response message contains the following information—basic and primary management CIDs assigned to this SS, RF power level adjustment, offset frequency adjustment, and timing offset. After receiving the ranging response from the BS, the SS will not transmit before it adjusts its RF signal in accordance with the ranging response. The SS must perform initial ranging at least once. Because operation in several channels must be supported, the subscriber station will attempt initial ranging on every suitable uplink channel before moving to the next available channel.

There are two types of ranging: initial ranging and periodic ranging. (Note that CDMA-based ranging is different.) After the initial ranging request, periodic ranging provides opportunities for other ranging requests and responses to adjust power levels, time, and frequency offsets. After the initial ranging request, further ranging requests by the SS can be transmitted using the basic CID. These ranging requests contain the power levels, time and frequency offsets. The BS will return another ranging response message to the SS with any additional fine-tuning required. These ranging request/response steps are repeated until the response contains a Ranging Successful notification, or the BS aborts ranging. Once successfully ranged (ranging request is within tolerance of the BS), the SS can transmit data in the uplink.

The BS provides each SS a periodic ranging opportunity. The period is BS dependent. When bandwidth is granted per subscriber station, any allocation of uplink bandwidth constitutes a ranging opportunity. The goal of this periodic ranging is to adjust the quality of the wireless link. The SS monitors the carrier to noise plus interference ratio $C/(N + I)$ and compares the average value against the allowed range of operation. If the received $C/(N + I)$ goes outside of the allowed operating region, the SS requests a change to a new burst profile. The SS applies an algorithm to determine its optimal burst profile. The channel descriptors (UCD and DCD) are updated regularly by the BS.

4) *Immediately after ranging is completed, the SS informs the BS of its basic capabilities.* The BS responds with a message containing the intersection of the SS's and the BS's capabilities.

5) *The BS and SS perform authorization and key exchange, as specified in the MAC security sublayer.* Each SS contains the following unique information when shipped from the manufacturer: a 48-bit IEEE-style MAC address and a X.509 digital certificate. These are assigned during the manufacturing process. The MAC address is used during the authentication and registration processes to establish the appropriate connections for SSs. The security information contained in the X.509 digital certificate is used to authenticate the SS to the security server and authenticate the responses from the security and provisioning servers.

6) *After authorization and key exchange, registration occurs.* The main goal of the registration process is for the SS to receive its secondary management connection ID and, therefore, to become manageable. In this process, the SS sends a registration request to the BS. The BS responds with a registration response message, which includes the secondary management connection ID. The registration response specifically authorizes the SS to forward traffic to the network.

7) *At this point, IP connectivity can be established on the SS's secondary management connection.* The SS uses standard DHCP mechanisms [B26] to obtain an IP address and other configuration parameters.

8) *The subscriber station and base station need to have the current date and time.* This is necessary for time-stamping certain events.

9) *The SS and the BS exchange operational parameters.*

10) *If there are any preprovisioned service flows in the service contract of this SS, the BS sets up connections for these preprovisioned service flows.*

Channel access and QoS

The principal mechanism for providing QoS is to associate packets with a service flow. The service flow, the central concept of the MAC protocol, is a unidirectional flow of packets with a particular QoS. The SS and BS provide this QoS according to the QoS parameter set defined for the service flow. Service flows can be preconfigured or dynamically established. The QoS parameters of a service flow can be specified implicitly by specifying a service class name or explicitly. If a service class name is given, modifying parameters can also be included. Service flows exist in both the uplink and downlink direction and may exist without actually being activated to carry traffic.

When the higher-layer data is forwarded to the MAC, the convergence sublayer includes the connection ID identifying the connection across which the information is to be transmitted. The service flow for the connection is mapped to the Connection Identifier (CID). The service class is an optional object that may be implemented at the BS. A service class is defined in the BS to have a particular QoS parameter set. The QoS parameter sets of a service flow may contain a reference to the service class name. The service flow QoS parameter sets may augment and even override the QoS parameter settings of the service class, subject to authorization by the BS.

The service class serves two purposes. First, it allows operators to move the burden of configuring service flows from the provisioning server to the BS. Second, it allows higher-layer protocols to create a service flow by its service class name. Service classes are merely identifiers for a specific set of QoS parameter set values. A service identified by a service class is treated no differently, once established, than a service that has the same QoS parameter set explicitly specified.

Service flows are distinguished by a 32-bit Service Flow Identifier (SFID); active service flows also have a 16-bit Connection Identifier (CID). A service flow is characterized by a set of QoS parameters such as bandwidth, latency, jitter, and throughput assurances.

To provide vendors with maximum flexibility, the way the QoS parameter set is provisioned is outside the scope of 802.16. Typically, the network management system will set up the QoS parameters. Following the centralized architecture, every BS must approve or deny every change to the QoS parameters. Within the BS, the authorization module approves or denies every change to the QoS parameters. This module performs the function of a QoS manager.

There is static and dynamic authorization. In the static authorization model, all possible services are defined in the initial configuration of each SS. The QoS parameters of provisioned service flows cannot be changed in the static model. Furthermore, the admitted QoS parameter set must be a subset of the provisioned QoS parameter set and the active QoS parameter set must be a subset of the admitted QoS parameter set.

In the dynamic authorization model, the authorization module also communicates through a separate interface to an independent policy server. This policy server may provide the authorization module with advance notice of upcoming admission and activation requests. The policy server also will specify the proper authorization action to be taken on those requests. Requests for admission and activation of service flows are checked by the authorization module to ensure that the QoS parameter set is a subset of the QoS set provided by the policy server. Before initial connection setup, the BS retrieves the provisioned QoS set for an SS. This is handed to the authorization module within the BS. The BS can implement mechanisms for overriding this automated approval process. Thus the dynamic model gives vendors the capability to implement any QoS policy desired. When modifying QoS parameters, care must be taken to prevent packet loss. In particular, any required bandwidth change is sequenced between the SS and BS. This sequencing means that if an uplink service flow's bandwidth is being reduced, the SS reduces its payload bandwidth first and then the BS reduces the bandwidth scheduled for the service flow. If an uplink service flow's bandwidth is being increased, the BS increases the bandwidth scheduled for the service flow first and then the SS increases its payload bandwidth.

It is useful to think about three basic types of service flows:

- *Provisioned.* A service flow may be provisioned but not immediately activated. Such a service flow can be referred to as "deferred." The BS assigns a Service Flow ID for such a service flow but does not reserve resources. As a result of external action typically at higher layers, the SS or the BS may choose to activate a provisioned service flow. Service flows will be provisioned by the network management system.

- *Admitted.* The protocol supports a two-phase activation model. Admitted Service Flows that have resources assigned but not yet completely activated are in a transient state. The goal is to use resources efficiently, to perform admission control quickly, and to prevent potential theft-of-service situations.

- *Active.* A service flow is an active service flow if it is requesting and being granted bandwidth for transporting data packets. This completes the two-stage activation model.

Service flows may be created, changed, or deleted. The provisioning of service flows is done by means outside of the scope of 802.16, such as the network management system. The creation of dynamic service flows may be initiated by the subscriber station or by the base station. This is accomplished through a series of MAC management messages. In addition, protocols are defined for modifying and deleting service flows. When a service flow is deleted, all resources associated with it are released. However, if a basic, primary management or secondary management service flow of an SS is deleted, that SS is deregistered and will have to reregister. Also, if a service flow that was provisioned during registration is deleted, the provisioning information for that service flow is lost until the SS reregisters. However, deleting a provisioned service flow shall not cause an SS to reregister. Therefore, care should be taken before deleting such service flows.

The MAC is connection-oriented. All data communications are in the context of a connection. Service flows can be provisioned when a subscriber station is installed. After subscriber station registration, connections are associated with these service flows, one connection per service flow. As customer's service needs change, new connections can be established or existing connections can be eliminated. The service flow defines the QoS parameters for the data that is exchanged on the connection. There are three connection management functions—creation, modification or maintenance, and deletion. The maintenance requirements vary depending upon the type of service connected. For example, connections that have a constant bandwidth requirement do not require maintenance. Connections that have dynamic (but relatively slowly changing) bandwidth requirements require some maintenance. IP services may require a substantial amount of ongoing maintenance due to their bursty nature and due to the high possibility of fragmentation. Existing connections may be changed in their characteristics on a dynamic basis in response to changing bandwidth requirements.

Uplink and downlink QoS is provided for each service flow.

The first parameter of QoS is bandwidth. A SS requests uplink bandwidth on a per-connection basis. However, the BS grants bandwidth on an aggregate basis for the entire SS. Requests for transmission are not associated with a subscriber station, but with a connection ID. Requests for transmission are always on a perconnection basis, because different connections can consume different amounts of bandwidth, even if they have the same service type. A subscriber station can have several connections. For example, a SS unit serving multiple tenants in an office building would make requests on behalf of all of them, though the service and other connection parameters may be different for each of them. Many higher-layer sessions may operate over the same wireless connection identifier. For example, many users within a company may be communicating with TCP/IP to different destinations, but bandwidth will be granted on an aggregate basis.

The main characteristic of the 802.16 standard is flexibility. This flexibility is extremely important for service providers, because it maximizes the revenue-generating potential of the technology. The multiple-access method enables the service to be tailored to the delay and bandwidth requirements of each user application. This flexibility is accomplished through five different types of uplink scheduling mechanisms. These are implemented using unsolicited bandwidth grants, polling, and contention procedures.

Mechanisms are defined in the protocol to allow vendors to optimize system performance using different combinations of these bandwidth allocation techniques. For example, contention may be used to avoid the individual polling of subscriber stations that have been inactive for a long period of time. The use of polling simplifies the access operation and guarantees that applications receive service on a deterministic basis if it is required. Scheduling services are designed to improve the efficiency of the poll/grant process. By specifying a scheduling service and its associated QoS parameters, the BS can anticipate the throughput and latency needs of the uplink traffic and provide polls and/or grants at the appropriate times. The basic services are Unsolicited Grant Service (UGS), real-time polling service (rtPS), non-real-time polling service (nrtPS) and Best Effort (BE) service. Each service is tailored to a specific type of data flow.

The bandwidth allocation and request mechanism are considered first. Increasing (or decreasing) bandwidth requirements is necessary for all services. Resources are assigned on a demand assignment basis as the need arises.

The Unsolicited Grant Service (UGS) is designed to support real-time service flows that generate fixed-size data packets on a periodic basis, such as T1/E1 and Voice over IP. The service offers fixed-size grants on a real-time periodic basis, which eliminates the overhead and latency of SS requests and assures that grants are available to meet the flow's real-time needs.

There are incompressible and compressible constant bit rate UGS connections. The needs of incompressible UGS connections do not change between connection establishment and termination. The requirements of compressible UGS connections may increase or decrease depending on traffic. In order for this service to work correctly, the subscriber station is prohibited from using any contention request opportunities and the BS is prohibited from providing any unicast request opportunities for that connection.

The real-time polling service (rtPS) is designed to support real-time data with variable-size packets on a periodic basis, such as MPEG video. The service offers real-time, periodic, unicast request opportunities, which meet the flow's real-time needs and allow the SS to specify the size of the desired grant. This service requires more request overhead than UGS but, because the grants are of variable size, the efficiency is higher. In order for this service to work correctly, the subscriber station is prohibited from using any contention request opportunities for that connection. Also, the BS must provide opportunities for periodic unicast requests even if a grant is pending. Consequently, the SS uses only unicast request opportunities to obtain uplink transmission opportunities. (The SS could still use unsolicited data grant burst types for uplink transmission as well.)

The non-real-time polling service (nrtPS) is designed to support non real-time data that require variable size data grant burst types on a regular basis, such as high bandwidth FTP. The service offers unicast polls on a regular basis, which assures that the flow receives request opportunities even during network congestion. The BS typically polls nrtPS connection identifiers on an interval on the order of one second or less. In order for this service to work correctly, the SS is allowed to use contention request opportunities. Therefore, the SS uses contention request opportunities, as well as unicast request opportunities and unsolicited data grant burst types.

The intent of the best effort (BE) service is to provide efficient service to best effort traffic. In order for this service to work correctly, the SS must be set to allow use of contention request opportunities. This results in the SS using contention request opportunities as well as unicast request opportunities and unsolicited data grant burst types.

Requests refer to the mechanism that SSs use to indicate to the BS that they need uplink bandwidth allocation. Note that, in the request intervals, collisions can happen. The BS controls assignments on the uplink channel and determines which minislots are request intervals and therefore subject to collisions. The mandatory method of contention resolution is based on truncated binary exponential backoff. The initial backoff window and the maximum backoff window are controlled by the BS. When an SS has information to send and wants to enter the contention resolution process, it randomly selects a number within its backoff window. According to the CSMA protocol, the SS defers transmission for a contention transmission opportunity equal to the selected backoff counter.

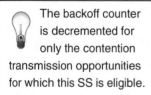
The backoff counter is decremented for only the contention transmission opportunities for which this SS is eligible.

This contention is only for the transmission of a bandwidth request. If the contention transmission is successful, the SS will receive a data grant. The subscriber station considers the contention transmission lost if no data grant has been given within a certain time interval. In this case, the SS

Chapter 4: Air Interface for Fixed Broadband Wireless Access Systems

will increase its backoff window by a factor of two, as long as it is less than the maximum backoff window. Then the SS will randomly select a number within its new backoff window and repeat the deferring process. This retry process continues until the maximum number of retries has been reached. The maximum number of retries is independent of the initial and maximum backoff windows.

Note that the contention resolution process can be abandoned. If the SS receives a data grant at any time while deferring, it will abandon the contention resolution process and use the explicit transmission opportunity. The centralized architecture of IEEE 802.16 manifests itself in giving the BS flexibility in controlling the contention resolution by dynamically adjusting the values of Backoff Start and Backoff End. At one extreme, the BS may choose to set up the backoff window to emulate an IEEE 802.11-style backoff. This will ensure fairness and efficiency. At the other end, the BS may make the Backoff Start and Backoff End identical so that all SS are using the same backoff window.

A request may come as a standalone bandwidth request, or it may come as a piggyback request. Note that because the uplink burst profile can change dynamically, the SS cannot estimate the amount of time that is required. This amount of time will depend on the modulation format that is used. Therefore, all requests for bandwidth are in terms of the number of bytes needed to carry the MAC header and payload, but not the PHY overhead. There are two types of distinguishable bandwidth requests—incremental or aggregate. When the BS receives an incremental bandwidth request, it will add the quantity of bandwidth requested to its current perception of the bandwidth needs of the connection. When the BS receives an aggregate bandwidth request, it will replace its perception of the bandwidth needs of the connection with the quantity of bandwidth requested. Piggybacked bandwidth requests are always incremental. The self-correcting nature of the request/grant protocol requires that SSs periodically use aggregate bandwidth requests. The period may be a function of the QoS of a service and of the link quality. Although bandwidth is requested for the individual connections, it is granted for the entire SS. This allows more intelligent SSs to implement dynamic bandwidth allocation mechanisms of their own. This may be useful for real-time applications that require a faster response from the system. Subscriber stations can use bandwidth stealing. A station may use bandwidth that has already been granted to send another request. This is called "bandwidth stealing."

Polling is the process by which the BS allocates to the SS's bandwidth specifically for the purpose of making bandwidth requests. Note that polling is done on either an SS or connection basis. Furthermore, individual subscriber stations or groups of subscriber stations can be polled. Unicast polling is when a subscriber station is polled individually. In this case, no explicit message is transmitted to poll the subscriber station. Rather, the SS is allocated sufficient bandwidth in the uplink map to respond with a bandwidth request.

If insufficient bandwidth is available to individually poll many inactive SSs, some SSs may be polled in multicast groups or a broadcast poll may be issued. Certain connections are reserved for multicast groups and for broadcast messages. As with individual polling, the poll is not an explicit message but bandwidth allocated in the uplink map. The difference is that, rather than associating allocated bandwidth with an SS's basic connection, the allocation is to a multicast or broadcast connections.

SSs with currently active UGS connections may set the poll-me bit in the grant management subheader to indicate to the BS that they need to be polled to request bandwidth for non-UGS connections; otherwise SSs are not polled during active UGS connections.

MAC SECURITY SUBLAYER

The security sublayer has two component protocols. The first is the encapsulation protocol for data encryption. This encapsulation protocol defines (1) a set of supported cryptographic suites, i.e., pairings of data encryption and authentication algorithms, and (2) the rules for applying those algorithms to a MAC payload. Authentication is based only on digital certificates. The second component protocol of the security sublayer is called privacy key management (PKM). The PKM protocol describes how the BS distributes keys to client SS in a secure fashion. Through this key management protocol, the SSs and the BS synchronize keys.

> The security, or privacy, sublayer is designed with two goals in mind. First, it provides subscribers with privacy across the fixed broadband wireless network. Second, it provides operators with strong protection from theft of service.

The set of security information that a base station and a subscriber station share is called security association. This security information includes traffic encryption keys and cipher block chaining (CBC) initialization vectors. Security associations are uniquely identified by a security association identifier (SAID). For connections that are mapped to a security association encryption, and data authentication are performed as specified by that security association. Three types of security associations are defined—primary, static, and dynamic.

Each SS establishes an exclusive primary security association with its BS during the SS initialization process. Static security associations are provided within the BS. Dynamic security associations are established and eliminated on the fly in response to the initiation and termination of specific service flows. Both static and dynamic SAs can be shared by multiple SSs. A security association's shared information includes the cryptographic suite employed within the SA. In this case, a cryptographic suite is the security association's set of methods for data encryption, data authentication, and traffic encryption key (TEK) exchange. The shared information may also include traffic encryption keys and initialization vectors. The exact content of the SA depends the SA's cryptographic suite. Security associations are identified using security association IDs. The ID of the primary security association of a SS must be equal to the basic connection ID of that SS.

 There are rules for mapping connections to security associations:
- All data connections are mapped to an existing security association.
- The basic and the primary management connections are not mapped to a security association, but the secondary management connection is mapped to the primary security association.
- All transport connections are mapped to a security association.

Encryption is always applied to the MAC payload; the header is not encrypted. The generic MAC header is not encrypted because it contains encryption information needed to decrypt a payload at the receiving station. There is a special bit field, called Encryption Control, which indicates whether the payload is encrypted. Any unencrypted MAC frame received on a connection mapped to a security association requiring encryption is discarded. To facilitate normal operation of the MAC sublayer, all MAC management messages are without encryption.

A SS uses the privacy key management (PKM) protocol to obtain authorization and traffic keying material from the BS, and to support periodic reauthorization and key refresh. The key management protocol uses X.509 digital certificates, the RSA public-key encryption algorithm, and strong symmetric algorithms to perform key exchanges between the SS and the BS. The privacy key management protocol adheres to a client/server model, where the SS is the client that requests keys, and the BS is the server that responds to those requests, ensuring that individual SS clients receive only keys for which they are authorized. The PKM protocol uses MAC management messaging. The key distribution mechanism has two steps. First, the PKM protocol uses public-key cryptography to establish a shared secret (i.e., an Authorization Key) between the SS and the BS. Then, the shared secret is used to secure subsequent exchanges of traffic encryption keys. This two-step mechanism for key distribution permits refreshing of traffic encryption keys without the significant overhead of computation-intensive public-key operations.

The keying material of a security association (e.g., DES key and CBC Initialization Vector) has a limited lifetime. When the BS delivers the keying material to an SS, it also provides the SS with that material's remaining lifetime. It is the responsibility of the SS to request new keying material from the BS before the current one expires at the BS. If the current keying material expires before a new set of keying material is received, the SS must perform network entry. Again, the PKM protocol specifies how the SS and the BS maintain key synchronization.

Using the PKM protocol, an SS requests from its BS the keying material for that security association. The BS ensures that each client SS only has access to the Security Associations it is authorized to access.

Authentication

The goal of the authentication process is for the BS to establish an authenticated identity of a client SS and the services (i.e., specific traffic encryption keys) the SS is authorized to access.

SS authorization, controlled by the authorization state machine, has three steps.

1) The BS authenticates a client SS's identity.

Chapter 4: Air Interface for Fixed Broadband Wireless Access Systems

2) The BS provides the authenticated SS with an authorization key, from which a Key Encryption Key (KEK) and message authentication keys are derived.

3) The BS provides the authenticated SS with the IDs of the security association and the properties of primary and static security associations for which the SS is authorized to obtain keys.

Each SS carries a unique X.509 digital certificate issued by the manufacturer itself or by an external authority. The digital certificate contains the SS's public key and SS MAC address. A SS begins authorization by sending an Authentication Information message to its BS. The Authentication Information message contains the SS's X.509 certificate. The Authentication Information message is strictly informative; i.e., the BS may choose to ignore it. However, it does provide a mechanism for a BS to learn the manufacturer certificates of its client SS. The SS sends an Authorization Request message to its BS immediately after sending the Authentication Information message. This is a request for an authorization key, as well as for the security association IDs identifying any static security associations in which the SS is authorized to participate. Again, the Authorization Request includes again the X.509 certificate, a description of the cryptographic algorithms the requesting SS supports, and the SS's basic connection ID. The SS's cryptographic capabilities are presented to the BS as a list of cryptographic suite identifiers, each indicating a particular pairing of encryption and authentication algorithms the SS supports. Recall that the basic connection ID is the first static CID the BS assigns to an SS during initial ranging—the primary security association ID is equal to the basic connection ID.

In response to an Authorization Request message, the BS verifies the digital certificate. The use of the X.509 certificates prevents cloned SSs from passing fake credentials onto a BS. The BS determines whether the requesting SS is authorized for basic unicast services and if the SS's user has subscribed to any additional statically provisioned services (i.e., static security association IDs). The BS associates the SS's authenticated identity to a paying subscriber, and hence to the data services that subscriber is authorized to access.

The protected services a BS makes available to a client SS can depend upon the particular cryptographic suites shared by the SS and the BS.

The BS then determines which encryption algorithm and protocol is common with the SS. From the list of cryptographic suites that the SS has provided, the BS selects one cryptographic suite to employ with the requesting SS's primary SA. The Authorization Reply that the BS sends back to the SS includes a primary security association descriptor, which, among other things, identifies the cryptographic suite the BS selected to use for the SS's primary SA. A BS rejects the authorization request if it determines that none of the offered cryptographic suites are satisfactory.

In addition to X.509, all SSs will either have factory-installed RSA private/public key pairs or provide an internal algorithm to generate such key pairs dynamically. If a SS relies on an internal algorithm to generate its RSA key pair, then the SS generates the key pair before its first authentication key (AK) exchange.

Initially the BS has a single active AK for each SS. After the BS has verified the digital certificate and selected a common cryptographic suite, it activates a second AK for this SS. The second AK is sent in the Authorization Reply message. The lifetime of this second Authorization key is finite and is set to be equal to the remaining lifetime of the first AK plus a predefined time period. In this fashion, for each authenticated client SS there are two active authorization keys with overlapping lifetimes. Then the BS activates an AK for the SS, uses the verified public key of the SS to encrypt this AK and then sends the encrypted key back to the requesting SS in an Authorization Reply message. The authorization reply includes the following:

1) An AK encrypted with the SS's public key
2) A four-bit key sequence number, used to distinguish between successive generations of authorization keys
3) A key lifetime
4) The identities (i.e., the IDs of the security association) and properties of the single primary and zero or more static security associations for which the SS is authorized to obtain keying information.

The Authorization Reply also contains an optional list of static security association descriptors; each static security association descriptor identifies the cryptographic suite employed within that security association. The selection of a static security association's cryptographic suite is typically made independent of the requesting SS's cryptographic capabilities. In its Authorization Reply, a BS may include static SA descriptors identifying cryptographic suites that the requesting SS does not support. This completes initial authorization.

An authorization request can be rejected. Generally, there are two types of rejections—permanent or simple. Permanent authorization rejection indicates that the error is of a permanent nature. What is interpreted as a permanent error is subject to administrative control within the BS. Examples of permanent error conditions include unknown manufacturer of the SS, invalid signature on the SS certificate, inconsistencies between data in the certificate and data in accompanying PKM data, or incompatible security capabilities. Simple authorization rejection does not indicate if the failure was due to a permanent error condition. As a result, the SS can set a wait timer and can reattempt authorization after the timer expires.

However, authorization is not a one-time process. Authorization is always valid for a finite period of time. After achieving initial authorization, a SS periodically seeks reauthorization with the BS. A SS must periodically refresh its authorization key by reissuing an Authorization Request to the BS. Reauthorization is identical to authorization with the exception that the SS does not send Authentication Information messages during reauthorization cycles. To avoid service interruptions during reauthorization, successive generations of the SS's authorization keys have overlapping lifetimes. Both SS and BS should be able to support up to two simultaneously active authorization keys during these transition periods. If the SS does not complete the reauthorization before the expiration of the authorization keys, the SS is considered unauthorized.

An SS must maintain its authorization status with the BS in order to be able to refresh aging TEKs. The refreshing of TEKs is managed by state machines. Upon achieving authorization, a SS starts a separate TEK state machine for each of the security association IDs identified in the Authorization Reply message. Each TEK state machine operating within the SS is responsible for managing the keys associated with its respective security association ID. TEK state machines periodically send Key Request messages to the BS, requesting a refresh of keying material for their respective security association identifier. The BS responds to a

Key Request with a Key Reply message, containing the BS's active keying material. The traffic encryption key (TEK) in the Key Reply is triple DES (encrypt-decrypt-encrypt or EDE mode) encrypted, using a two-key, triple DES key encryption key (KEK) derived from the authorization key.

> A SS must maintain the two most recent generations of keys for each SA. This retention helps maintain uninterrupted service during the key transitions.

Since every key associated with a Security Association (SA) has a limited lifetime, the BS manages a two-bit key sequence number independently for each SA and distributes this key sequence number along with the SA's key to client SS. The BS increments the key sequence number with each new generation of keys. The MAC header includes this sequence number to identify the specific generation of the SA key used to encrypt the attached payload. Because the sequence number is a two-bit quantity, it wraps around to 0 when it reaches 3. Comparing a received MAC frame's key sequence number with what it believes to be the "current" key sequence number, an SS or BS can easily recognize a loss of key synchronization with its peer.

As discussed previously, at all times the BS maintains two active sets of keys per security association identifier. The lifetimes of the two generations overlap such that each generation becomes active halfway through the life of its predecessor and expires halfway through the life of its successor. A BS includes in its Key Replies both of the SAID's active generations of keying material. The Key Reply provides the requesting SS, in addition to the TEK and CBC initialization vectors, their remaining lifetimes. The receiving SS uses these remaining lifetimes to estimate when the BS will invalidate a particular TEK, and therefore when to schedule future Key Requests so that the SS requests and receives new keying material before its current keying material expires. The operation of the TEK state machine's Key Request scheduling algorithm, combined with the BS's regimen for updating and using an SAID's keying material, ensures that the SS will be able to continually exchange encrypted traffic with the BS.

A TEK state machine remains active as long as the SS is authorized to operate in the BS's security domain, i.e., it has a valid authorization key, and the SS is authorized to participate in that particular security association, i.e., BS continues to provide fresh keying material during rekey cycles.

There are two state machines—authorization state machine and TEK state machine. The authorization state machine will start an independent TEK state machine for each of its authorized SAIDs. The BS maintains two active TEKs per SAID. The BS includes in its Key Replies both of these TEKs, along with their remaining lifetimes. The BS encrypts downlink traffic with the older of its two TEKs. Note that the BS encrypts with a given TEK for only the second half of its lifetime. The BS decrypts uplink traffic with either the older or newer TEK, depending upon which of the two keys the SS was using at the time. The SS encrypts uplink traffic with the newer of its two TEKs and decrypts downlink traffic with either the older or newer TEK, depending upon which of the two keys the BS was using at the time.

Communication between authorization and TEK state machines occurs through the passing of events and protocol messaging. The authorization state machine generates events (i.e., Stop, Authorized, Authorization Pending, and Authorization Complete events) that are targeted at its child TEK state machines. TEK state machines do not target events at their parent authorization state machine. The TEK state machine affects the authorization state machine indirectly through the messaging a BS sends in response to a SS's requests.

Through operation of a TEK state machine, the SS attempts to keep its copies of the SAID's TEKs synchronized with those of its BS. A TEK state machine issues Key Requests to refresh copies of its SAID's keying material soon after the scheduled expiration time of the older of its two TEKs and before the expiration of its newer TEK. To accommodate for SS/BS clock mismatch and timing mismatches, the SS schedules its Key Requests several seconds before the newer TEK's estimated expiration in the BS. When it receives the Key Reply, the SS updates its records with the TEK Parameters from both TEKs contained in the Key Reply message.

The BS is responsible for maintaining keying information for all Security Associations. The BS must maintain two active TEKs and their associated initialization vectors per Security Association identifier.

Data encryption with DES

The encryption method used is the Cipher Block Chaining (CBC) mode of the U.S. Data Encryption Standard (DES). The CBC initialization vector is calculated as follows. In the downlink, the CBC is initialized with the XOR of the initialization vector in the TEK keying information and the PHY synchronization field of the latest DL-MAP. In the uplink, the CBC is initialized with the XOR of the initialization vector in the TEK keying information and the PHY synchronization field of the DL-MAP that is in effect when the UL-MAP is received.

In the Key Reply message, TEK is encrypted by the BS using the two-key triple-DES in the encrypt-decrypt-encrypt (EDE) mode. This encryption operation can be described by

$$C = Ek1[Dk2[Ek1[P]]] \qquad \text{Eq. 4–1}$$

where P is 64-bit plaintext TEK, C is 64-bit ciphertext TEK, $E[]$ is 56-bit DES electronic codebook (ECB) mode encryption, and $D[]$ is 56-bit DES ECB decryption. The leftmost 64 bits of the 128-bit KEK are denoted $k1$, and the rightmost 64 bits of the KEK are denoted $k2$. The corresponding decryption operation is

$$P = Dk1[Ek2[Dk1[C]]] \qquad \text{Eq. 4–2}$$

The 56-bit DES keys are actually 64 bits long. In every byte, the seven most significant bits are independent and are used in a DES key. The least significant bit is a parity bit, so that every byte has odd parity.

The keying material for two-key triple DES consists of two distinct DES keys. The 64 most significant bits of the KEK are used in the encrypt operation. The 64 least significant bits are used in the decrypt operation.

MAC ENHANCEMENTS FOR 2–11 GHZ OPERATION

There is a significant interest in the wireless industry especially in the frequency bands between 2 GHz and 11 GHz. The advantages for operation in this frequency range include operation in a non-line of sight (NLOS) environment. Operation in this frequency band requires not only a different physical layer, but also additions (or, alternatively, enhancements) to the MAC layer. These enhancements to the MAC are necessary to support not only point-to-multipoint architectures but, in license-exempt bands, optional mesh topology. Because multipath is significant, losses over the wireless medium will increase. To deal with these losses, the second area of additions to the MAC are related to introduction of ARQ on a per-connection basis. The third area of additions is related to the DFS and TPC mechanisms.

MAC enhancements for mesh systems

In unlicensed bands the topology of the network can be different from the point-to-multipoint type. In particular multipoint-to-multipoint (or mesh) networks can be used. Mesh architecture is a network where devices are capable of forwarding traffic from and to multiple other devices. Figure 4–5 is an illustration of the mesh concept.

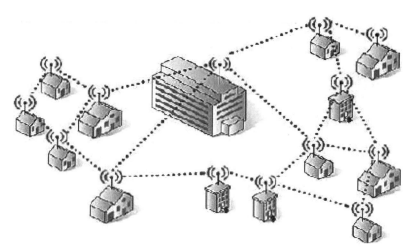

Figure 4–5: An example of a mesh network

Mesh mode is optional for operation in unlicensed bands between 2 GHz and 11 GHz. In mesh networks, a number of subscriber stations within a geographic area are interconnected and can act as repeater stations. This allows a variety of routes between the core network and any subscriber station. Clearly, at least one of the devices in a mesh network must be connected to a backhaul network. This device is termed a mesh BS. Devices that are not connected to the backhaul network are termed mesh SS. Uplink and downlink are different for mesh systems compared with point-to-multipoint networks. Downlink is defined as traffic away from the mesh BS, and uplink is defined as traffic towards the mesh BS. The other three important elements of the architecture of mesh systems are neighbor, neighborhood, and extended neighborhood. The stations with which a node has direct links are called neighbors. All neighbors of a node form a neighborhood. The neighbors of a node are considered to be "one hop" away from the node. An extended neighborhood also contains all the neighbors of the neighborhood.

Mesh networks need a different multiple-access mechanism than point-to-multipoint systems. The reason is that mesh devices that are close to each other cannot transmit at the same time on the same channel. If a node is transmitting all nodes within a distance of two hops must remain silent. Therefore, even the BS must coordinate its transmission with other stations. The transmission activity of a node depends substantially on how much traffic it will have to forward to and from other nodes. In turn, the amount of traffic depends on the position of the node within the network. The main difference between mesh and point-to-multipoint mode is that the channel resources (e.g., the ability to transmit) are shared between the systems on demand basis. There are two principal ways to assign channel resources: distributed scheduling and centralized scheduling. A combination of both is also possible. Distributed scheduling implies distributed decision-making. In centralized scheduling, the mesh BS makes decisions as to who has the right to transmit at any given time instant.

 Clearly, mesh networks need changes in the basic MAC protocol because of their architecture. The mesh capability support includes (1) mesh network configuration message, (2) mesh network entry message, and (3) messages for distributed and centralized mesh scheduling.

Mesh devices use mostly roof-mounted omnidirectional antennas, because coverage must be extended in all directions. However, mesh systems can use a variety of antenna types simultaneously. At the edge of the coverage area of the mesh network even highly directional antennae can be used.

In the coordinated distributed scheduling mode, all stations (BSs and SSs) coordinate their transmissions in their extended two-hop neighborhood. Every station (BS and SS) transmits a message at regular intervals to inform all the neighbors of its own schedule and proposed schedule changes. This message is used simultaneously to convey resource requests to the neighbors. All the stations in a network must use this same channel to transmit schedule information in a format of specific resource requests and grants. Coordinated distributed scheduling ensures that transmissions are scheduled in a manner that does not rely on the operation of a BS, and that are not necessarily directed to or from the BS. Within the constraints of the coordinated schedules (distributed or centralized), uncoordinated distributed scheduling can be used for fast, ad-hoc setup of schedules on a link-by-link basis. Uncoordinated distributed schedules are established by directed requests and grants between two nodes These schedules must be set up to keep the resulting data transmissions (and the request and grant packets themselves) from colliding with the data and control traffic scheduled by the coordinated distributed methods or the centralized scheduling methods. The differences between the coordinated and uncoordinated distributed scheduling are that, in coordinated scheduling, management messages are scheduled in the control subframe in a collision-free manner. In uncoordinated scheduling, management messages can collide.

Centralized scheduling is performed by the BS. A Mesh Centralized Scheduling Configuration message is broadcasted in mesh systems using centralized scheduling. The BS shall broadcast this message to all its neighbors, and all the subscriber stations forward (rebroadcast) the message in a way specified in the message. The mesh BS gathers resource requests from all the mesh SSs within a certain threshold hop range (HR). The mesh BS will determine the amount of granted resources for each link in the network both in downlink and uplink, and communicates these grants to all the mesh SSs within the hop range. The grant messages do not contain the actual schedule, but each node can compute it by using the predetermined algorithm with given parameters.

The network connections and topology are the same as in the distributed scheduling mode. However, some portion of the scheduled transmissions for the SSs less than or equal to a certain number of hops from the BS will be either defined by the BS or computed by the SSs themselves. This hop range, HR threshold, may be determined at the system startup phase or may be dynamic, according to considerations such as network density, the proximity of other BSs, and/or the dynamic characteristics of the traffic streams.

In the basic form, the BS provides the schedule for all the SSs less than or equal to HR threshold hops from the BS. The BS determines the flow assignments from the resource requests from the SSs within the HR threshold hop range. Subsequently, the SSs themselves determine the actual schedule from these flow assignments by using a common algorithm that divides the frame proportionally to the assignments. Thus, the BS acts just like the BS in a PMP network except that not all the SSs have to be directly connected to the BS and the assignments determined by the BS extends also to those SSs not directly connected to the BS but less than HR threshold hops from it. The SS resource requests and the BS assignments are both transmitted during the control portion of the frame. Centralized scheduling ensures that transmissions are coordinated to ensure collision-free scheduling over the links in the routing tree to and from the BS. Centralized scheduling is more efficient than the distributed scheduling for traffic streams (or collections of traffic streams that share links) that persist over a duration greater than the cycle time to relay the new resource requests and distribute the updated schedule.

In the absence of clearly defined uplink and downlink, only TDD is supported in mesh mode.

The first portion of a downlink (MAC) frame contains unencrypted frame control information destined for all SSs. This control section must contain a DL-MAP message for the downlink channel followed by one UL-MAP message for each associated uplink channel. The DL-MAP message does not describe the locations of the payload bursts that immediately follow; it describes the locations of the downlink bursts at some prescribed time in the future. This delay is necessary because of MAC and PHY processing delays.

Since all the basic functions like scheduling and network synchronization are based on the neighbor information, each node (BS and SS) in the mesh network maintains a physical neighborhood list containing the MAC address and the distance (measured in hops) of this neighbor from the present node. A SS that has a direct link to the BS will synchronize to the BS, while a SS that is at least two hops from the BS will synchronize to its neighbor SSs that are closer to the BS. Special packets, called network configuration and network entry packets, provide a basic level of communication between nodes in different nearby networks. These packets are used to synchronize both centralized and distributed control mesh networks. These packets are also used to synchronize the transmissions among multiple co-located BSs.

Node initialization and network entry procedures in mesh mode are in some aspects different from those in PMP mode. A new node entering the mesh network obeys the following procedure. First, it goes through scanning and synchronization. After scanning and synchronization, a station listens for mesh configuration packets to build its physical neighbor table. Then it selects a sponsor node from the neighbor table based on receive signal quality, and gains a coarse network synchronization using a received network configuration packet from the selected sponsor. Then the new node transmits a packet, which includes the selected sponsor node's address. After a sponsor node is selected, the new station can perform authentication. The sponsor node will forward authentication frames to a node in the mesh, called authentication node. This authentication node will act as a BS, verifying the digital certificate of the new station. MAC management messages are forwarded to the new station through its sponsor node. A node can exchange TEKs with other nodes, not just the BS. A TEK state machine is identified by a SAID. Every node will maintain separate TEK state machines per SAID for all nodes it exchanges TEKs with. If operation with a selected sponsor fails, a different sponsor is selected and the procedure repeated. The sponsor will also include the link information for this new node in its neighbor list. After authorization the new station receives its node ID, establishes IP connectivity, time of day, and exchanges the operational parameters.

Advanced antenna systems (AAS)

Systems that work in the 2–11 GHz band can use advanced antenna systems (AAS). Advanced antenna systems are based on a technology that uses more than one antenna element. A significant advantage of this technology is that it improves range and system capacity. These are very important advantages, especially when due to a combination of technical and regulatory reasons that cannot be easily achieved in other ways. It is expected that advanced antenna systems will be used in the future, and 802.16 as an option includes functionality to support them.

The first advantage of AAS is that their spectral efficiency can be increased linearly with the number of antenna elements. This is achieved by steering beams to multiple users simultaneously so as to realize an intercell frequency reuse of one and an in-cell reuse factor proportional to the number of antenna elements. An additional benefit is the SNR gain realized by coherently combining multiple signals. Another possible benefit is the reduction in interference achieved by steering nulls in the direction of co-channel interferers. It must be noted that these techniques do not require multiple antennas at the SS. However, it is still possible to use multiple antennas at the SS to achieve further benefits.

Initially only the BS will have AAS. The design of the AAS option provides a mechanism to migrate from a non-AAS system to an AAS-enabled system, in which the initial replacement of the non-AAS-capable BS by an AAS-capable BS should not cause service interruption. The goal of 802.16 is to make possible systems where the BS, and possibly some SS, use AAS, while other SSs do not. This is achieved by dedicating only part of any frame to AAS traffic (Figure 4–6). Part of the frame is dedicated to non-AAS traffic and part is dedicated to AAS traffic. The allocation is performed dynamically by the BS. Non-AAS SSs can ignore AAS traffic, which they can identify based on the DL-MAP/UL-MAP messages. AAS-enabled SSs will use dedicated DL-MAP/ UL-MAP messages and are therefore prevented from colliding with non-AAS traffic. Non-AAS subscriber stations are hence able to operate whether or not they receive any signal in the AAS part of the frame, and AAS subscriber stations are able to operate whether or not they receive any signal in the non-AAS part of the frame.

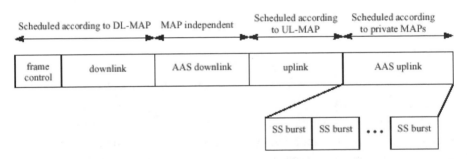

Figure 4–6: Frame structure for systems with advanced antennas

Additional functionality required to support AAS in the MAC layer can be divided into two categories—control functionality and utility functionality. The control functionality is responsible for the DL-MAP and UL-MAP distribution, and channel description. The utility function is responsible for the PHY-related information provided by the MAC.

Furthermore the following additional logical channels are required for AAS support:

1) *Downlink Synchronization Channel:* This channel is used for time and frequency synchronization by AAS subscriber stations.

2) *Downlink Polling Channel:* This channel is required for systems where the array adaptation using signals received in the Uplink Contention Channel may not be adequate to open a Downlink Traffic Channel to a subscriber station.

3) *Downlink Traffic Channels:* This is the part of the physical channel that carries downlink user traffic, as scheduled by the BS. Their availability requires array adaptation. In contrast to similar channels in the non-AAS systems, multiple simultaneous Downlink Traffic Channels can be open to spatially separated subscriber stations.

4) *Uplink Contention Channel:* This channel is used for SS-initiated random access, including ranging and bandwidth request. Its availability does not require array adaptation.

5) *Uplink Traffic Channels:* This is the part of the physical channel that carries uplink user traffic, as scheduled by the BS. Their availability requires array adaptation.

The process of registration in AAS systems is different from the regular registration process. This is because the adaptive array operating in the PHY cannot be effective until the MAC and PHY of the BS identify the registering subscriber station. The adaptation of BS antenna array can be accomplished only after the BS has identified the SS. On the other hand, if the BS cannot adapt the array to the new subscriber, there is a chance that the SS will not receive any valid signal from the BS at all. Because the regular registration process requires that the SS receive a valid message first, it is not always possible to rely on this process. If the SS decodes the DL-MAP and DCD messages, it can proceed with the network entry process, just like in the non-AAS case (power adjustment, rate adaptation, authentication, encryption, etc.). In this case, the BS can adapt its adaptive array during the ranging process.

> Note also that AAS subscribers might not be able to request bandwidth using the usual contention mechanism. This happens because the adaptive array may not have a beam directed at the SS when it is requesting BW, and the BW request will be lost. There are two ways out of this situation. The SS can broadcast the bandwidth request or the BS can poll the SS.

If the SS cannot decode the DL-MAP and DCD messages, it can use the following procedure: An AAS BS must reserve a fixed, predefined part of the frame as initial-ranging contention slots for AAS SS. These contention slots are located at a well-known location relative to the downlink preamble, so even an SS that can only identify the DL preamble can locate it. These contention slots are called AAS-alert-slots and their number and location in the frame is PHY specific. The AAS SS that cannot decode the DL-MAP and DCD messages will attempt ranging on the AAS-alert-slots. Then the SS will wait for a private DL-MAP and private DCD messages from the BS. If these messages are received, the SS continues the network entry process like a non-AAS SS. If the private DL-MAP and DCD messages are not received, the SS will use an exponential backoff algorithm for selecting the next frame in which to attempt alerting the BS for its presence.

Adaptive Arrays require knowledge of the channel at both downlink and uplink. The BS can obtain channel information in two ways. The first way is to perform uplink channel estimation and assume that the downlink channels are identical to the uplink channels. The second way is to use feedback from the SS, thus transmitting the estimated channel state from the SS to BS. The first method is simpler and is well suited for TDD systems. The second method is more suitable for FDD systems, where reciprocity does not apply (because uplink and downlink are in different frequency bands). The second method can also be used in TDD systems. In the second method, the BS must periodically send a message to the SS to signal that channel state information should be updated. The period with which this message is sent is called Channel Estimation Interval (CEI). It is vendor-specific and is an internal parameter of the BS. a. The value of CEI is defined according to the channel stability over time (a typical value is 20 ms). The BS is responsible to determine the actual value of CEI, and for the distribution of this value to all SSs. The SS must perform channel estimations on a regular time basis to be able to provide up-to-date estimations upon request. In response to a request from the BS, the SS will reply with the most up-to-date estimation of the channel.

Automatic repeat request (ARQ)

Multipath in the range between 2 GHz and 11 GHz leads to intersymbol interference, reduced signal-to-noise ratio, and increased BER and PER. To deal with these significant issues, 802.16 provides an ARQ mechanism. The ARQ mechanism is an optional part of the MAC layer and can be enabled on a per-connection basis. The per-connection ARQ and associated parameters are specified and negotiated during connection creation or change. A connection cannot have a mixture of ARQ and non-ARQ traffic. Similar to other properties of the MAC protocol, the scope of a specific instance of ARQ is limited to one unidirectional connection. The ARQ feedback information can be sent as a standalone MAC management message on the appropriate basic management connection or piggybacked on an existing connection. ARQ feedback cannot be fragmented. The receiver provides positive or negative ARQ acknowledgments.

Recall that the format of a MAC frame is a single header and one or more whole or fragmented payloads. In addition to the single MAC header, there may be subheaders. The MAC header contains information whether subheaders are present. If ARQ is used, a MAC frame can contain ARQ feedback payload, preceded by an appropriate subheader.

Transmitter and receiver state machine operations include comparing fragment sequence numbers and taking actions based on which is larger or smaller. In this context, it is not possible to compare the numeric sequence number values directly to make this determination. The ARQ protocol is responsible for choosing the right fragment size on a per-fragment basis. The fragment size is not a fixed value that remains constant for a particular connection. Once defined, the size of a fragment cannot be changed between retransmissions. An ARQ fragment may be in one of the following four states—not-sent, outstanding, discarded, and waiting-for-retransmission. Any ARQ fragment begins as not-sent. After it is sent, it becomes outstanding for a period of time termed the retry timeout. While a fragment is in outstanding state, either it is acknowledged and then discarded, or transitions to state waiting-for-retransmission after a timeout. An ARQ fragment can become waiting-for-retransmission before the timeout expires if it is negatively acknowledged. An ARQ fragment may also change from waiting-for-retransmission to discarded when a positive acknowledgment for it is received or after another timeout expires. The transmitter policy is that if any waiting-for-retransmission ARQ fragments exist, they should be given precedence over not-sent packets for the same connection. When fragments are retransmitted, the fragment with the lowest frame sequence number is retransmitted first. The ARQ transmit fragment state sequence is shown in Figure 4–7.

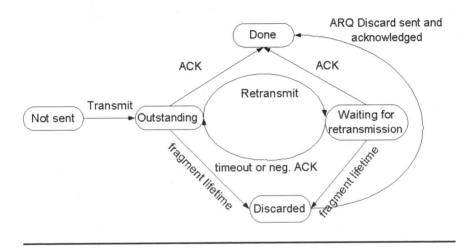

Figure 4–7: ARQ transmitter state machine

When an acknowledgment is received, the transmitter checks the validity of the frame sequence number. To be valid, a frame sequence number should be in a specified range. If the frame sequence number is not valid, the transmitter will ignore the acknowledgment. There is cumulative acknowledgment for multiple fragments. The cumulative acknowledgment contains a bitmap. All fragments where the corresponding bit is 1 in the bitmap are positively acknowledged, and those with the corresponding bit set to 0 are negatively acknowledged.

Under abnormal error conditions, as a last resort, a station may send a Reset message. Exactly when a Reset is sent is outside the scope of 802.16.

When a frame is received, its integrity is determined based on the CRC-32 checksum. If the frame passes the checksum, it is unpacked and defragmented, if necessary. The receiver maintains a sliding window. The start of the window is the lowest fragment that has not been received or has been received in error. When an ARQ fragment with a number in the range defined by the sliding window is received, the receiver will accept it. ARQ fragment numbers outside the sliding window are rejected as out of order.

For each ARQ fragment accepted fully and without errors, an acknowledgment message is sent to the transmitter. Acknowledgments may be either for specific ARQ fragments (i.e., contain information on the acknowledged ARQ fragment numbers), or cumulative (i.e., contain the highest ARQ fragment number below which all ARQ fragments have been received correctly) or a combination of both (i.e., cumulative with selective). Acknowledgments are sent in the order of the ARQ fragment numbers they acknowledge. How often acknowledgments are generated depends on implementation. The interval is not specified in 802.16. The ARQ receiver state machine has four states—not received, received, received correctly, and done. Initially a fragment is not received. If the CRC check is correct and the frame sequence number is in range a fragment is received correctly. If the fragment sequence number is out of range, the fragment is considered not received.

DFS for license-exempt operation

For operation in unlicensed bands, DFS is required. Unlicensed bands can be allocated for more than one use. In addition to being available on unlicensed basis, some bands have primary users. These primary users of the bands legally have the right of way. Wireless data communication systems can operate in unlicensed bands only if they are capable of avoiding the primary users by employing a dynamic frequency selection (DFS). If a frequency band contains a primary user, the wireless data communications equipment must detect the primary user and move to another band. Devices can move to another band for other reasons such as low RSSI or interference from other systems. Therefore, DFS is used when devices choose frequency band for operation. Each downlink channel will be characterized in terms of its RSSI. Valid channels are only those channels that have a high enough SNR. DFS procedures provide for the following:

- Requesting and reporting measurements
- Testing channels for the presence of primary users
- Detecting primary users
- Ceasing operation on a channel after primary users have been found
- Selecting and advertising a new channel

Note that several of these procedures are outside of the scope of 802.16.

The process also requires channel state monitoring by the SS. Channels are assigned by the BS. The initial ranging mechanism consists of a series of transactions where the SS send a ranging request message, and the BS answers with a ranging response message and the SS corrects its transmission power or timing parameters as instructed by the BS. During periodic maintenance, the SS can be either polled by the BS to transmit a ranging request message, or can send it independently. The DFS mechanism messages are appended to the ranging messages. In the ranging response message, the BS can request the DFS info. The SS shall reply with an appropriate ranging request message containing the requested information. The ranging request/response transactions continue until the BS is satisfied with the results.

MAC enhancements for OFDM and OFDMA PHYs

The WirelessMAN-OFDM physical layer supports two contention-based bandwidth request mechanisms. In addition to the mandatory bandwidth request mechanism, the MAC provides another bandwidth request mechanism. The station requesting bandwidth can send what is called a contention code. This contention code is modulated on a contention channel. The contention channel consists of four carriers. The contention code and the contention channel are randomly chosen with uniform distribution. When the BS receives this contention code' it will provide an predefined UL allocation for the SS to transmit a bandwidth request and, optionally, additional data. If the BS does not provide an uplink allocation, the SS will assume that the ranging code transmission resulted in collision and will follow the contention resolution procedure.

The OFDMA physical layer supports two mandatory contention-based bandwidth request mechanisms. In addition to the standard mechanism for bandwidth requests, there is a CDMA-based mechanism. The OFDMA layer specifies a Ranging Subchannel and a set of Ranging codes. These ranging codes are special PN codes used for making bandwidth requests. The SS selects with equal probability a ranging code from the code subset. The selected ranging code is modulated onto the ranging subchannel and transmitted during the appropriate uplink allocation. When the BS receives this contention code, it will provide a predefined uplink allocation for the SS. If the BS does not provide an uplink allocation, the SS will assume that the ranging code transmission resulted in collision and will follow the contention resolution procedure.

IEEE 802.16 PHYSICAL LAYERS

Every physical layer has two sublayers—a transmission convergence sublayer and a Physical Medium Dependent (PMD) sublayer. Each PMD sublayer may require the definition of a unique transmission convergence sublayer. In addition, a physical layer may be accompanied by a physical layer management entity. IEEE 802 standards do not specify the exact functions of the management entity; the functions are always left to vendors. In general, the management entity is responsible for such functions as gathering of layer-dependent status and interaction with the general system management. Physical layer management functions include frequency adjustment, power management, propagation delay compensation, etc. The management information specific to PHY layer is represented as a management information base (MIB) for this layer. The generic model of the device management is that it either reads or sets the value of a parameter in the MIB.

Recalling the architecture of IEEE 802.16 shown in Figure 4–2, the PHY can be considered as providing services to the MAC layer. These services are accessed through the PHY service access point (SAP). Therefore the communication between the MAC and the PHY is described using a set of primitives. There are three types of primitives between the MAC and the PHY—primitives for data transfer, primitives for management functions, and service primitives that have local significance and support sublayer-to-sublayer interactions related to layer control.

IEEE 802.16 defines several physical layers, for business and other reasons. The main reason is that there was not significant support within the working group for a single physical layer. By standardizing several physical layers at once, vendors are left free to implement whichever they want. Ultimately this flexibility allows service providers the ability to optimize system deployments with respect to cell planning, cost, radio capabilities, services, and capacity.

Physical layer for 10–66 GHz

The characteristics of outdoor wireless channels are different. In outdoor wireless communication at frequencies over 10 GHz, the multipath components will be absorbed rather than reflected. The energy of the non-line-of-sight components that reach the receiver will be very small. This is a typical scenario for the operation of wireless Internet access equipment. The IEEE 802.16 working group has studied the characteristics of wireless channels and has come up with three types of channels.

The first type of channel is type 0, where no multipath components are present and the frequency response of the channel is equal to a constant. Channel type 1 has one weak non-line-of-sight component, and its frequency response can be expressed as:

$$H(j\omega) = 0.995 + 0.0995 e^{-j0.75} e^{-j\omega 400/B} \qquad \text{Eq. 4–3}$$

where B is a parameter that can assume values between 15 and 25. This parameter is the baud rate that is used for communication over this channel, measured in Mbaud.

Channel type 2 has the strongest multipath components. Its frequency response can be expressed as:

$$H(j\omega) = 0.286 e^{-j0.75} + 0.953 e^{-j\omega 400/B} + 0.095 e^{-j\omega 800/B} \qquad \text{Eq. 4–4}$$

At frequencies between 10 GHz and 66 GHz, the predominant fade mechanism in outdoor wireless communication is that resulting from rain attenuation. Given certain parameters such as frequency of operation, desired link availability and BER, the rain attenuation will determine the maximum cell radius of operation. An internationally accepted method for computation of rain fade attenuation probability is that defined by ITU-R 530 [B92]. As an example, for operation at 28 GHz and link availability of 99.995%, the maximum cell radius is about 3.5 km in ITU rain region *K*.

A significant mitigation technique for the control of both intrasystem and intersystem interference is the antenna radiation pattern envelope (RPE) discrimination provided by system antennas. The RPE discrimination has significance for both clear sky and rain faded propagation conditions. The RPE requirements for aggressive intrasystem frequency reuse plans may exceed the RPE requirements for the control of intersystem coexistence. Recommended antenna RPE characteristics are described in "Coexistence " on page 317.

> Rain attenuation will be different in general for the desired signal and for the interfering signal, i.e., there will be a differential rain fading loss. This differential rain fading loss between the two transmission paths can have a significant impact on the system performance. For example, at operational frequencies around 28 GHz, the estimated rain cell diameter is approximately 2.4 km.

The physical layer for operation between 10 GHz and 66 GHz must allow for flexible spectrum usage and support both time division duplexing (TDD) and frequency division duplexing (FDD). The wireless channel is different for the different SSs. The burst transmission format of the PHY is framed so that it supports adaptive burst profiling in which transmission parameters, including the modulation and coding schemes, can be adjusted individually to each subscriber station (SS) on a frame-by-frame basis. The FDD case supports full duplex SSs as well as half duplex SSs, which do not transmit and receive simultaneously. Note that because the architecture of IEEE 802.16 is substantially different from the architecture of IEEE 802.11 and IEEE 802.15, it uses adaptive modulation. Adaptive modulation means that systems communicate using multiple burst profiles, and furthermore the BS communicates with multiple subscriber stations using different burst profiles.

The uplink physical layer is based on a combination of time division multiple access (TDMA) and demand-assigned multiple access (DAMA). In particular, according to the TDMA part the uplink channel is divided into a number of time slots. The number of slots assigned for various uses (registration, contention, guard, or user traffic) is controlled by the MAC layer in the base station and may vary over time for optimal performance.

The downlink channel is time division multiplexed (TDM), with the information for each subscriber station multiplexed onto a single stream of data and received by all subscriber stations within the same sector. To support half duplex FDD subscriber stations, a provision is also made for a TDMA portion of the downlink. The downlink physical layer includes a transmission convergence sublayer that inserts a pointer byte at the beginning of the payload to help the receiver identify the beginning of a MAC frame. The PMD layer performs randomization, FEC encoding, and modulation according to QPSK, 16-QAM, or 64-QAM (optional).

The uplink physical layer is based upon TDMA burst transmission. Each burst is designed to carry variable-length MAC frames. The PMD layer in the uplink performs randomization of the incoming data, FEC encoding, and modulation using QPSK, 16-QAM (optional), or 64-QAM (optional) The parameters the 10–66 GHz PHY are summarized in Table 4–1.

Table 4–1: Parameter values for the 10–66 GHz PHY

Parameter	Value
Symbol rate	16 to 40 Mbaud
Modulation density	2, 4, or 6
FEC block size	0–511 bytes
FEC payload	0–255 bytes
Uplink preamble length	16 or 32 symbols
PHY overhead	0–256 symbols

Power can adjusted in steps of 0.5 dB, k 0.5, where $k = -128...127$.

This physical layer specification operates in a framed format. The supported frame durations are 0.5 ms, 1 ms, and 2 ms. Each frame contains a downlink subframe and an uplink subframe. The downlink subframe begins with information necessary for frame synchronization and control. In the TDD case, the downlink subframe comes first, followed by the uplink subframe. In the FDD case, uplink transmissions occur concurrently with the downlink frame. Each SS attempts to receive all portions of the downlink except for those bursts whose burst profile is either not implemented by the SS or is less robust than the SS's current operational downlink burst profile. Half duplex SSs do not listen to portions of the downlink coincident with their allocated uplink transmission.

In FDD operation, the uplink and downlink channels are on separate frequencies. The capability of the downlink to be transmitted in bursts facilitates the use of different modulation types and allows the system to simultaneously support full duplex subscriber stations (which can transmit and receive simultaneously) and half duplex subscriber stations (which do not).

In the case of TDD, the uplink and downlink transmissions share the same frequency but are separated in time. A TDD frame also has a fixed duration and contains one downlink and one uplink subframe. The TDD framing is adaptive in that the link capacity allocated to the downlink versus the uplink may vary. Between the downlink burst and the subsequent uplink burst there is a gap, referred to as the Tx/Rx transition gap. This gap allows time for the BS to switch from transmit to receive mode and SSs to switch from receive to transmit mode. Similarly, there is a gap between the uplink burst and the subsequent downlink burst, called the Rx/Tx transition gap. Both transition gaps are an integer number of physical slots durations and start on a physical slot boundary.

Downlink physical layer

The available bandwidth in the downlink direction is defined in terms of physical slots. The available bandwidth in the uplink direction is defined in terms of mini-slots, where the minislot length is $2m$ physical slots *(m ranges from 0 through 7)*. The number of physical slots with each frame is a function of the symbol rate. The symbol rate is selected in order to obtain an integral number of physical slots within each frame. For example, with a 20 Mbaud symbol rate, there are 5000 physical slots within a 1 ms frame.

The downlink frames can be TDD and FDD. The TDD downlink subframe contains preamble, frame control section, and data. The preamble is used for synchronization and equalization. The frame control section contains downlink and uplink maps stating the physical slots at which bursts begin. The frame control section is not encrypted and is destined for all SSs. It also contains information about the modulation type and FEC code length for each burst profile. The data is organized into bursts with different burst profiles and therefore different level of transmission robustness. The bursts are transmitted in order of decreasing robustness. For example, with the use of a single FEC type with fixed parameters, data begins with QPSK modulation, followed by 16-QAM, followed by 64-QAM. At the end of the frame there is a Tx/Rx transition gap. Each SS receives and decodes the control information of the downlink and looks for MAC headers indicating that the remainder of the downlink subframe contains data for that SS.

In the FDD case, similar to the TDD case, the downlink subframe begins with a preamble followed by a frame control section and a TDM portion organized into bursts transmitted in decreasing order of burst profile robustness. This TDM portion of the downlink subframe contains data transmitted to full duplex SSs, half duplex SSs scheduled to transmit later in the frame than they receive, and half duplex SSs not scheduled to transmit in this frame. The FDD downlink subframe continues with a TDMA portion used to transmit data to any half duplex SSs scheduled to transmit earlier in the frame than they receive. This allows an individual SS to decode a specific portion of the downlink without the need to decode the entire downlink subframe. In the TDMA portion, each burst begins with the downlink TDMA burst preamble for phase resynchronization. Bursts in the TDMA portion need not be ordered by burst profile robustness. The FDD frame control section includes a map of both the TDM and TDMA bursts.

The TDD downlink subframe, which inherently contains data transmitted to SSs that transmit later in the frame than they receive, is identical in structure to the FDD downlink subframe for a frame in which no half duplex SSs are scheduled to transmit before they receive.

The frame control section contains the DL-MAP message, followed by the UL-MAP message for each associated uplink channel. In addition, it may contain DCD and UCD messages.

Two downlink burst preambles are used. The Frame Start Preamble is at the beginning of each downlink frame. The Downlink TDMA Burst Preamble is at the beginning of each TDMA burst in the TDMA portion of the downlink

Chapter 4: Air Interface for Fixed Broadband Wireless Access Systems

subframe. Both preambles use QPSK modulation and are based upon +45 degrees rotated constant amplitude zero autocorrelation (CAZAC) sequences. These sequences can be normalized, because the amplitude of the preamble depends on the downlink power adjustment rule. In the case of the constant-peak power scheme, the preamble is transmitted so that its constellation points coincide with the modulation scheme's outermost constellation points. In the case of the constant mean power scheme, the preamble is transmitted with the mean power of the constellation points of the modulation scheme in use.

A burst corresponds to either a TDM burst beginning with the Frame Start Preamble or a TDMA burst beginning with a Downlink TDMA Burst Preamble.

The Frame Start Preamble consists of a 32-symbol sequence generated by repeating the 16-symbol QPSK CAZAC sequence $1 + j, -1 + j, -1 - j, 1 - j, 1 + j, -1 - j, 1 + j, -1 - j, 1 + j, 1 - j, -1 - j, -1 + j, 1 + j, 1 + j, 1 + j, 1 + j$. The Downlink TDMA Burst Preamble consists of a 16-symbol sequence generated by repeating the 8-symbol CAZAC sequence $1 + j, 1 + j, 1 + j, -1 + j, -1 - j, 1 + j, -1 - j, -1 + j$. In the case of TDMA downlink, a burst includes the Downlink TDMA Burst Preamble of length p physical slots, and the downlink map entry points to its beginning.

The first portion of the message after the preamble is the frame control section. The frame control section consists of a DL-MAP message followed by an UL-MAP message. The DL-MAP message contains information at which physical slots the burst profile changes. In general, the number of physical slots i (which is an integer) allocated to a particular burst can be calculated from the downlink MAP, which indicates the starting position of each burst as well as the burst profiles. In the frame control section, there may be DCD and UCD messages. The DCD message contains information about the burst profile, such as modulation type and FEC code type and parameters.

In the TDM portion, downlink data is transmitted in order of decreasing burst robustness. In the case of TDMA, data are grouped into bursts, but are not necessarily grouped in the order of robustness.

The downlink physical medium dependent layer is summarized in the block diagram in Figure 4–8.

Chapter 4: Air Interface for Fixed Broadband Wireless Access Systems

The downlink channel supports adaptive burst profiling on the user data portion of the frame. Up to 12 burst profiles can be defined. The parameters of each are communicated to the SSs via MAC messages during the frame control section of the downlink frame.

Because there are optional modulation and FEC schemes that can be implemented at the subscriber station, a method for identifying the capability to the base station is required (i.e., including the highest order modulation supported, the optional FEC coding schemes supported, and the minimum shortened last codeword length supported). This information is communicated to the base station during the subscriber registration period.

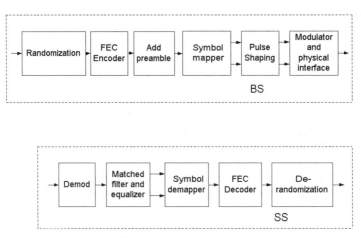

Figure 4-8: Block diagram of the downlink physical medium dependent sublayer

The operation of each block in Figure 4–8 will be described next.

The first signal-processing operation that the transmitted data is subject to is randomization. Randomization prevents long sequences of ones or zeros. An adequate number of bit transitions facilitates clock recovery. The downlink data is randomized by modulo-2 addition (alternatively, XOR) of every data bit with the output of a pseudorandom binary sequence generator, specified as a shift register. The preambles are not randomized. The randomizer sequence is applied only to information bits. The generator polynomial for the shift register is $x^{15} + x^{14} + 1$

and the initialization sequence is 100101010000000. The rightmost bit has weight 15, the next one has weight 14, and so on. At the beginning of each burst, the shift register is cleared and initialized. The first randomization bit is the modulo-2 addition of the two rightmost bits of the initialization sequence. Clearly, the first six bits of the randomization sequence are all zeros.

For the same business and technical reasons that were outlined in the introduction of this chapter, four forward error correction (FEC) schemes are defined in 802.16. Defining four FEC schemes allows the performance of IEEE 802.16 equipment to improve over time, together with improvements in semiconductor and signal processing technologies. Notably these improvements do not require going through the standardization process again.

The 802.16 standard defines concatenated coding schemes. Concatenated coding has been studied for a long time and is well known. Concatenated coding methods employ two constituent codes: an inner code and an outer code. The reason for using two types of codes is that in general there are two types of errors—random errors and burst errors.

In code type 1, the outer code is a systematic Reed-Solomon code over $GF(256)$ and no inner code is used. The block length is $N = 255$ bytes, and the information block length K varies from 6 to 255 bytes. As a result, this code is able to correct errors ranging from 16 to 0 bytes. Therefore, it can be represented as RS $(255, 255 - R)$ where R is the number of redundancy bytes and ranges between 0 and 32, inclusive. The value of K is a parameter specified for each burst profile by the MAC. Code type 1 is useful either for a large data block or when high coding rate is required. This code has two modes of operation—fixed codeword and shortened last codeword operation.

In fixed codeword operation, the number of information bytes K is the same in each Reed–Solomon codeword. If the MAC messages in a burst require fewer bytes than are carried by an integral number of codewords, stuff bytes (FF in hexadecimal notation) are added so that the total message length is an integral multiple of K bytes. The SS and the BS can determine the number of codewords in its downlink burst. Using the burst length, both the SS and the BS calculate the number of full-length RS codewords that can be carried by each burst.

In the shortened last codeword mode of operation, the number of information bytes in the final Reed–Solomon block of each burst is reduced from the normal number K, while the number of parity bytes R remains the same. The BS reduces the number of information bytes in the last codeword to minimize the number of stuff bytes to add to the end of the MAC message. The length of the burst is then set to the minimum number of physical slots required to transport all of the burst's bytes, which include preamble, information, and parity bytes. The BS implicitly communicates the number of bytes in the shortened last codeword to the SS via the Downlink Map message, which defines the starting physical slot of each burst. The SS uses the Downlink Map information to calculate the number of full-length RS codewords and the length of the shortened last codeword that can be carried within the specified burst size. The BS performs a similar calculation as the SS for its encoding purposes. To allow the receiving hardware to decode the previous Reed–Solomon codeword, in this mode of operation a codeword will have at least 6 information bytes. The number of information bytes carried by the shortened last codeword is between 6 and K bytes, inclusive. Only if the number of information bytes to be sent by the BS is less than 6 bytes will stuff bytes be added at the end.

In code type 2 the outer code is the same RS code as in code type 1, with one minor restriction. When using code type 2 in the shortened last codeword mode of operation, the number of information bytes in the shortened last codeword must be an even number. Consequently, if an odd number of information bytes must be sent, a stuff byte is added because the inner code requires a pair of bytes on which to operate.

The inner code is a (24,16) block convolutional code (BCC). This inner code in code type 2 can be considered as an equivalent nonsystematic block code. The input to the block code is 16 bits, and the output (the codeword) is 24 bits long. The encoder of this equivalent block code can be described by the following equations:

$c23 = b15 \oplus b0 \oplus b1$

$c22 = b15 \oplus b1$

$c21 = b14 \oplus b15 \oplus b0$

$c20 = b13 \oplus b14 \oplus b15$

$c_{19} = b_{13} \oplus b_{15}$

$c_{18} = b_{12} \oplus b_{13} \oplus b_{14}$

$c_{17} = b_{11} \oplus b_{12} \oplus b_{13}$

$c_{16} = b_{11} \oplus b_{13}$

$c_{15} = b_{10} \oplus b_{11} \oplus b_{12}$

$c_{14} = b_9 \oplus b_{10} \oplus b_{11}$

$c_{13} = b_9 \oplus b_{11}$

$c_{12} = b_8 \oplus b_9 \oplus b_{10}$

$c_{11} = b_7 \oplus b_8 \oplus b_9$

$c_{10} = b_7 \oplus b_9$

$c_9 = b_6 \oplus b_7 \oplus b_8$

$c_8 = b_5 \oplus b_6 \oplus b_7$

$c_7 = b_5 \oplus b_7$

$c_6 = b_4 \oplus b_5 \oplus b_6$

$c_5 = b_3 \oplus b_4 \oplus b_5$

$c_4 = b_3 \oplus b_5$

$c_3 = b_2 \oplus b_3 \oplus b_4$

$c_2 = b_1 \oplus b_2 \oplus b_3$

$c_1 = b_1 \oplus b_3$

$c_0 = b_0 \oplus b_1 \oplus b_2.$

Code type 2 is useful for low-to-moderate coding rates that provide good performance.

While code types 1 and 2 must be implemented by all BSs and SSs, code types 3 and 4 are optional. In code type 3, the outer code is still the same, but the inner code is (9, 8) parity check code. The code itself is a simple bit-wise parity check.

This code adds one parity bit to every eight bits. The parity bit is the sum modulo 2 of all eight bits.

In code type 4, the outer code is a block turbo code (BTC), and no inner code is used. A block turbo code has the properties of a concatenated code. Code type 4 has significant advantages. It can be used to either extend the range of a base station or increase the data rate at the same range. The BTC idea is illustrated in Figure 4–9.

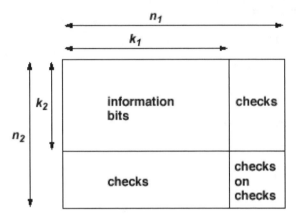

Figure 4–9: Two-dimensional product code matrix

Essentially, the idea of turbo codes is to encode the data twice. First, the k_1 information bits in the rows are encoded into n_1 bits, by using an (n_1,k_1) block code. Then the k_2 information bits in the columns are encoded into n_2 bits by using the same or possibly different block code. The parity bits of the first code are also encoded by the second code. The overall block size of such a product code is $n = n_1 \times n_2$, the total number of information bits is $k = k_1 \times k_2$, and the code rate is k/n. As a constituent block coding scheme the standard specifies (31,26) and (63,57) extended Hamming block codes. Shorter codes can be obtained by deleting an appropriate number of rows and columns.

When using the Block Turbo Coding, bit interleaving can be used. At present, two modes of interleaving are defined. The first mode of interleaving, type 1, means actually no interleaving. In this mode, the encoded bits are read from the encoder row by row, in the order that they were written. Type 2 interleaver is a block interleaver. In this mode, the encoded bits are read from the encoder after the first

k_2 rows are written into the encoder memory. The bits are read column by column, proceeding from the top position in the first column. Currently, a type 3 interleaver is not defined. It is possible that in the future an interleaver that has advantages in terms of performance and/or implementation can be defined.

It is important that the frame control section of the downlink frame be encoded with a fixed set of parameters known to the SS at initialization in order to ensure that all subscriber stations can read the information. The control portion of the frame is encoded with a type 2 FEC, where the outer code is a (46, 26) Reed–Solomon code and an inner code is a (24, 16) BCC. Following FEC, QPSK modulation is used.

In communication systems, digital modulation specifies how bits are mapped to symbols. The downlink physical layer of IEEE 802.16 supports several modulation formats. The constellation can be selected per subscriber based on the quality of the RF channel. In a high SNR environment, a more complex modulation scheme can be used. If the air-link degrades, the system can revert to the less complex constellations. The goal is to maintain a constant error rate. In the downlink, the mandatory modulation formats are QPSK and 16-QAM. 64-QAM is an option. Figures 4–10 through 4–12 describe the bit mappings for QPSK, 16-QAM, and 64-QAM modulation.

The distance from the origin determines the power that sends the signal. There are two power adjustment rules—constant constellation peak power and constant constellation mean power. In the constant peak power scheme, corner points are transmitted at equal power levels regardless of modulation type. In the constant mean power scheme, the signal is transmitted at equal mean power levels regardless of modulation type. The power adjustment rule is specified in the DCD management message.

Prior modulation, the I and Q signals are filtered by industry-standard square-root raised cosine pulse-shaping filters with excess bandwidth factor of 0.25. Finally, at the antenna port the transmitted waveform is given by:

$$S(t) = I(t)\cos(2\pi f_c t) - Q(t)\sin(2\pi f_c t) \qquad \text{Eq. 4–5}$$

where $I(t)$ and $Q(t)$ are the signals after pulse-shaping, and f_c is the RF carrier frequency.

Chapter 4: Air Interface for Fixed Broadband Wireless Access Systems

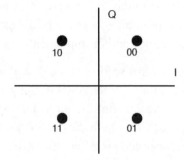

Figure 4–10: Bit-to-symbol mapping for QPSK in the downlink

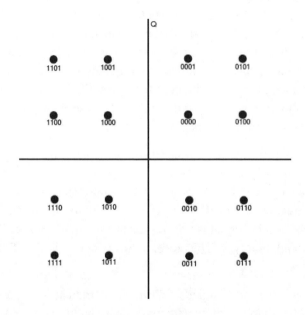

Figure 4–11: Gray-coded 16-QAM in the downlink

Chapter 4: Air Interface for Fixed Broadband Wireless Access Systems

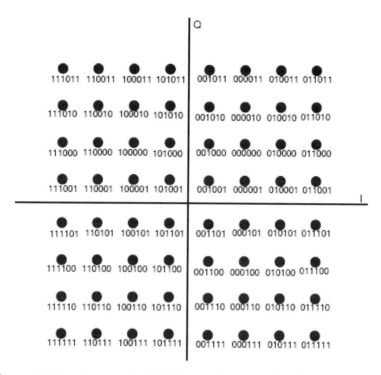

Figure 4–12: Gray-coded bit-to-symbol mapping for the optional 64 QAM in the downlink

Uplink physical layer

Three classes of bursts may be transmitted by the SS during the uplink subframe—bursts transmitted in contention opportunities reserved for initial maintenance, bursts transmitted in contention opportunities provided by multicast and broadcast polls, and bursts transmitted in intervals specifically allocated to individual SSs. One frame can contain any of these bursts. Bursts can occur in any order and in any quantity (limited by the number of available physical slots) within the frame, at the discretion of the BS uplink scheduler. All uplink transmissions are made according to the uplink burst profiles, specified by the BS, as indicated in the UL-MAP message in the frame control section of the downlink.

The burst profiles include information such as burst profile, FEC type, and preamble length.

SS transition gaps separate the transmissions of the various SSs during the uplink subframe. The gap allows some finite time for the previous burst to subside, followed by a preamble allowing the BS to synchronize to the new SS. The preamble and gap lengths are broadcast periodically in the UCD message.

Each uplink burst begins with an uplink preamble. This preamble is based upon a repetition of a +45 degrees rotated constant amplitude zero autocorrelation (CAZAC) sequence. Two preambles can be used—one of 16-symbol length or one of 32-symbol length. The base station defines the preamble length and communicates it to the SS in the UCD message. The 16-symbol preamble is formed by repeating twice the following CAZAC sequence:

$$\{1+j,-1+j,1+j,1+j,-1-j,-1+j,-1-j,1+j\}.$$

The 32-symbol preamble is formed by repeating twice

$$\{1+j,1+j,1-j,-1-j,-1-j,1+j,1+j,-1+j,1+j,1+j,-1+j,1+j,-1-j,1+j,-1-j,1-j\}$$

These are QPSK sequences. The amplitude of the preamble depends on the uplink power adjustment rule—constant peak power or constant mean power. In the case of constant peak power, the constellation points of the preamble coincide with the outermost constellation points of the modulation scheme in use. In the case of the constant mean power, the mean power of the constellation points of the preamble is the same as the mean power of the modulation scheme in use.

The uplink transmission convergence sublayer is identical to the downlink transmission convergence sublayer. The uplink physical medium dependent layer coding and modulation are summarized in the block diagram shown in Figure 4–13.

Randomization and FEC encoding in the uplink are identical to the corresponding operations in the downlink.

The type of modulation and the power adjustment rule are set by the base station and are communicated in the UCD message. QPSK is mandatory, and the 16-QAM and 64-QAM are optional. The mappings of bits to symbols are identical to those in the downlink. An SS should not produce an EIRP spectral density exceeding either +30 dBW/MHz or the local regulatory requirements.

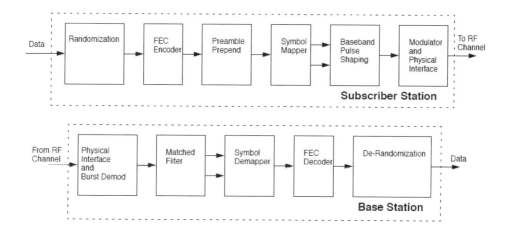

Figure 4–13:   Uplink PHY

The band between 10 GHz and 66 GHz is very wide, and regulatory requirements vary significantly between different regions. Due to wide variations in local regulations, no frequency plan is specified in 802.16. No single plan can accommodate all cases. To ensure that products built to 802.16 have interoperability over the air interface, the parameters in Table 4–2 should be followed. Systems shall use Nyquist square-root raised cosine pulse shaping with a roll-off factor of 0.25. Note that baud rates are chosen to provide an integer number of physical slots per frame. A frame duration of 1 ms is recommended as a compromise between transport efficiency (with lower frame overhead) and latency.

Note that for 802.16, synchronization and carrier frequency offset estimation and compensation, among the subscriber stations and the BS are more important than in the other wireless standards. Following the centralized architecture, the BS will make periodic measurements of the carrier frequency offset and will communicate it to the subscriber stations by a MAC message.

Table 4–2: Baud rates and channel sizes for a roll-off factor of 0.25

Channel size (MHz)	Symbol rate (Mbaud)	Bit rate (Mb/s) QPSK	Bit rate (Mb/s) 16 QAM	Bit rate (Mb/s) 64 QAM
20	16	32	64	96
25	20	40	80	120
28	22.4	44.8	89.6	134.4

An important part of the physical layer that is left to vendors is the power control algorithm. To provide power control in the uplink, the base station must be capable of providing accurate power measurements of the received burst signal. Then the BS will send information about the power level back to the subscriber stations in calibration messages. The power control algorithm should also take into account various power fluctuations due to the characteristics of the transmission medium, the performance of the RF power amplifier in different burst profiles, and other factors. The exact implementation of power control is vendor-specific.

Physical layers for 2–11 GHz operation

The physical layer in the previous section is for the 10–66 GHz band. Operating in licensed bands requires a license, which is very expensive, if at all possible to obtain. IEEE 802.16a specifies a physical layer for operation at frequencies between 2 GHz and 11 GHz. Clearly the band between 2 GHz and 11 GHz also includes unlicensed bands. The protocol for operation in the unlicensed frequencies is called WirelessHUMAN. The 2–11 GHz bands provide a physical environment where, due to the longer wavelength, line of sight is not necessary and multipath may be significant. The channel bandwidths used in this physical environment typically vary from 1.5 MHz to 14 MHz. The ability to support non-line-of-sight requires additional PHY functionality, such as the support of advanced power management techniques, interference mitigation/coexistence, and smart antennas.

Both single-carrier and multiple-carrier modulation can be used. In general, the goal of a modulation technique is to transfer data over a prescribed channel bandwidth, within transmit power, reliability, and receiver complexity constraints.

Single-carrier systems require an equalizer at the receiver to compensate for any distortion resulting from the channel, because the multipath creates intersymbol interference among the received symbols. This equalization can be performed in time, frequency, and mixed (time and frequency) domain. In multicarrier systems, data is mapped to symbols and then multiplexed using an orthogonal transformation into a number of simultaneous lower-speed streams. The discretetime signal after the orthogonal transformation is converted to an analog waveform and mixed to a RF frequency. The nonlinear distortions introduced as part of the RF conversion can create significant out-of-band interference and must be reviewed in the context of:

- Deployment in specific frequency bands and out-of-band requirements,
- Coexistence with adjacent (in-band and out-of-band) systems (in particular TDD/FDD),
- Trade-offs in terms of guard bands, system performance, and system complexity.

In multicarrier systems, the total available channel bandwidth is subdivided among these multiple streams. Although the frequency response over the entire channel bandwidth may not be ideal (i.e., nonconstant), the spacing between the modulated carriers is small, so that the frequency response over the signaling bandwidth of one carrier is approximately constant. Equalization for multicarrier systems is still required, but it becomes a simple multiplication per carrier.

IEEE 802.16's specification for unlicensed bands (known as WirelessHUMAN for Wireless High-Speed Unlicensed Metropolitan Area Networks) is specifically designed for operation in the 5 GHz band, but it can be applicable to other license-exempt bands in the 2–11 GHz range. In the U.S. and Canada, the 5.15–5.25 GHz lower U-NII band is restricted to indoor use only. Because in BWA the BS is outdoors, this band is not available for BWA.

Operation in the unlicensed bands introduces additional interference and coexistence issues. Government regulations limit the allowed radiated power. IEEE 802.16a introduces mechanisms to detect and avoid interference such as dynamic frequency selection (DFS). Here again, DFS means the ability of a system to switch to different physical RF channels based on channel measurement criteria. DFS is a feature that all wireless technologies operating in unlicensed bands must have.

All frequency layers between 2 GHz and 11 GHz have some common characteristics. Frequency division duplex (FDD), half duplex frequency division (H-FDD), and time division duplex (TDD) modes provide for bidirectional operation, except for operation in the license-exempt band, where provision is made for TDD operation only. TDD flexibility permits efficient allocation of the available bandwidth and hence is capable of efficiently allocating the available traffic transport capacity for applications with varying ratios of uplink-to-downlink traffic transport demand. TDD operates in single, paired, or noncontiguous blocks of frequencies. FDD/H-FDD can be used by applications that require fixed asymmetric allocation between their uplink and downlink traffic transport demand. FDD and H-FDD operate in paired downlink/uplink sub-bands.

Table 4–3 gives a summary of the different BWA systems.

Table 4–3: BWA systems

Standard	Targeted frequency band	PHY	MAC	Duplexing
WirelessMAN-SC	10–66 GHz	SC	Basic	TDD, FDD
WirelessMAN-SC2	2–11 GHz	SC2	Basic+ARQ+STC+AAS	TDD, FDD
WirelessMAN-OFDM	2–11 GHz licensed	OFDM	Basic+ARQ+STC+DFS+AAS	TDD, FDD
WirelessMAN-OFDM	2–11 GHz unlicensed	OFDM	Basic+ARQ+STC+DFS+mesh+AAS	TDD
WirelessMAN-OFDMA	2–11 GHz licensed	OFDMA	Basic+ARQ+STC+DFS+AAS	TDD, FDD
WirelessMAN-OFDMA	2–11 GHz unlicensed	OFDMA	Basic+ARQ+STC+DFS+mesh+AAS	TDD

WirelessMAN-SC2 Physical layer for 2–11 GHz frequencies

This physical layer is based on single-carrier technology and designed for NLOS operation in the 2–11 GHz band. To enable NLOS operation, this physical layer supports channel estimation and equalization. The main characteristics of this physical layer are:

- TDMA uplink
- TDM downlink
- Concatenated FEC using Reed-Solomon and pragmatic TCM codes
- Optional BTC and convolutional TC
- No FEC option using ARQ, optional STC
- Optional AAS support

The WirelessMAN-SC2 is similar to the 10–66 GHz physical layer. As shown in Figure 4–14, the first step is randomization (or scrambling), performed as in WirelessMAN-SC. Data randomization is performed on the source bits, before FEC encoding, on both the downlink (DL) and uplink (UL).

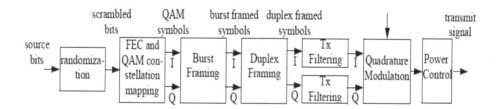

Figure 4–14: **WirelessMAN-SC2 transmitter signal processing**

The bits from the randomizer are passed to the channel encoder. The FEC mechanism is different from WirelessMAN-SC, because it is designed for a multipath environment. It is well-known that the performance of codes in channels with frequency selective fading is different from the performance in AWGN channels. Note that in general the amount of coding gain required may differ between uplink and downlink due to their different propagation characteristics.

In WirelessMAN-SC2, a concatenated FEC scheme is used, based on the serial concatenation of a Reed-Solomon outer code and a rate-compatible TCM inner code, as shown in Figure 4–15. Between the outer and the inner code, byte interleaving can be used as an option. The outer code is RS (255, 239) over GF (256). This code can be shortened and punctured to enable variable block sizes and variable error-correction capability. Shortening is when the number of information bytes that are encoded, K, is less than 239. In this case, the first $239-K$ data bytes of the block are set to zero but are not transmitted. When the data cannot be divided into an integer number of K-byte blocks, a shortened RS code will be used for the last block. The capability to shorten the code is mandatory, while puncturing is optional. Puncturing is when the code has fewer than 16 parity bytes. Puncturing may be used, depending on the burst profile.

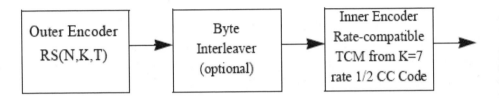

Figure 4–15: Concatenated coding scheme

The inner code is a rate-compatible pragmatic TCM code, derived from the rate 1/2 constraint length $K = 7$, binary convolutional code, which is identical to the convolutional code in IEEE 802.11a, with generator polynomials designated by 171 and 133 in octal notation. In addition to rate 1/2, this convolutional code can also support other rates through puncturing. Through puncturing, rates of 2/3, Ω, 5/6, and 7/8 can be obtained. Table 4–4 summarizes the mandatory and optional modulations and coding rates for WirelessMAN-SC2.

The pragmatic TCM code is constructed from both nonsystematic coded bits taken from the rate 1/2 binary convolutional encoder outputs and systematic uncoded bits taken directly from the encoder input. The resulting coded bits are then mapped to symbol constellations. The choice of a particular code rate and modulation is made based on burst profile parameters.

Table 4–4: Mandatory (M) and optional (O) modulations and coding rates for WirelessMAN-SC2

Modulation	Supported coding rates	Support	
		Uplink	Downlink
BPSK	1/2, 3/4	M	O
QPSK	1/2, 2/3, 3/4, 5/6, 7/8	M	M
16-QAM	1/2, 3/4	M	M
64-QAM	2/3, 5/6	O	M
256-QAM	3/4, 7/8	O	O

Interleaving between the inner and outer code is optional. Interleaving is not used in broadcast burst profiles. When interleaving is used, its usage and parameters will be specified within a burst profile. The interleaver changes the order of bytes from the Reed Solomon (RS) encoder output. Interleaving is used to spread consecutive bits into separate symbols after modulation. In this way, the interleaver prevents a series of consecutive bad symbols, which may occur due to multipath. A de-interleaver in the receiver restores the order of the bytes before RS decoding. The interleaver is a block interleaver, where a table is written a byte at a time row-wise (one row per RS code word) and read a byte at a time column-wise. The number of columns is equal to the codeword length of the Reed–Solomon code. The number of rows used by the interleaver is a burst parameter and is equal to the number of RS codewords that the interleaver can store. So that bursts exceeding an intended receiver's capabilities are not generated, the largest number of rows supported by a station is communicated during SS basic capability negotiation. The product of the number of rows and the number of columns is the interleaver block size.

The bits exiting the inner code (after puncturing, if it is used) are mapped to I and Q symbol coordinates using either a pragmatic TCM or Gray code symbol map, depending on the code rate and modulation scheme. All BPSK and QPSK code rates and rate 1/2 16-QAM shall use the Gray-coded constellation maps depicted in Figure 4–22 on page 297. Rate 3/4 16-QAM and all code rates for 64-QAM use the pragmatic TCM constellation map depicted in Figure 4–23 on page 297. All code rates for 256-QAM use the pragmatic TCM constellation map depicted in Figure 4–24 on page 298.

Figure 4–16 illustrates the rate 1/2 pragmatic TCM encoder for 16-QAM. The baseline rate 1/2 binary convolutional encoder first generates a two-bit constellation index, $b_3 b_2$, associated with the I symbol coordinate. Then it generates a two-bit constellation index, $b_1 b_0$, for the Q symbol coordinate. The I index generation precedes the Q index generation. Note that this encoder should be interpreted as a rate 2/4 encoder, because it generates one four-bit code symbol per two input bits. For this reason, input bit sequence that feeds this encoder must have length divisible by two.

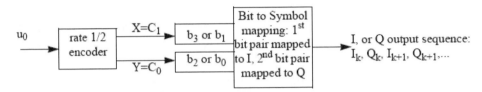

Figure 4–16:   TCM encoder for rate 1/2 16 QAM

Figure 4–17 illustrates the rate 3/4 pragmatic TCM encoder for 16-QAM. This encoder uses the same rate 1/2 binary convolutional encoder, along with two systematic bits that are passed directly from the encoder input to the encoder output. With this structure, the encoder is capable of simultaneously generating four output bits per three input bits. The sequence of arrival for the $u_2 u_1 u_0$ input into the encoder is u_2 arrives first, u_1 second, u_0 last. During the encoding process, the encoder generates a two-bit constellation index, ($b_3 b_2$, for the I symbol coordinate) and simultaneously generates another two bit constellation index, designated $b_1 b_0$, for the Q symbol coordinate. The length of the input bit sequence must be divisible by three. Figure 4–23 depicts the bits-to-symbol-constellation map that is applied to the rate 3/4 16-QAM encoder output.

Chapter 4: Air Interface for Fixed Broadband Wireless Access Systems

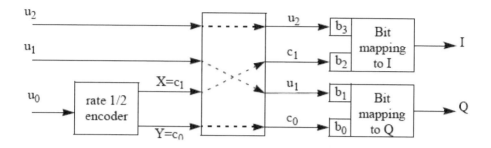

Figure 4–17: Rate 3/4 pragmatic TCM encoder for 16 QAM

Figure 4–18 illustrates the rate 2/3 pragmatic TCM encoder for 64-QAM. This encoder uses again the same rate 1/2 binary convolutional encoder, along with one systematic bit that is passed directly from the encoder input to the encoder output. The sequence of arrival for the $u_1 u_0$ input into the encoder is u_1 arrives first, u_0 last. The encoder (as a whole) then generates a three bit constellation index, $b_5 b_4 b_3$, which is associated with the I symbol coordinate. From the next two bits, the encoder generates another three-bit constellation index, $b_2 b_1 b_0$, which is associated with the Q symbol coordinate. The I index generation should precede the Q index generation. Figure 4–23 depicts the bits-to-symbol-constellation map that is applied to the rate 2/3 64-QAM encoder output.

> This encoder should be interpreted as a rate 4/6 encoder, because it generates one six-bit code symbol per four input bits. For this reason, the length of the input bit sequence must be divisible by four.

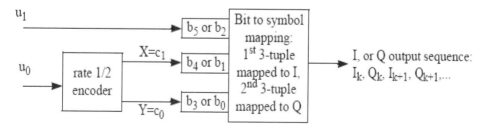

Figure 4–18: Pragmatic TCM encoder for rate 2/3 64 QAM

Wireless Communication Standards 293

Figure 4–19 illustrates the rate 5/6 pragmatic TCM encoder for 64-QAM. This encoder uses a rate 3/4 punctured version of the rate 1/2 binary convolutional encoder, along with two systematic bits that are passed directly from the encoder input to the encoder output. The rate Ω punctured code is generated from the baseline rate 1/2 code using an appropriate puncture mask. Puncture samples are sequenced c_3 first, c_2 second, c_1 third, and c_0 last. The sequence of arrival for the $u_4u_3u_2u_1u_0$ input into the encoder is u_4 arrives first, u_3 arrives second, u_2 arrives third, u_1 arrives next to last, and u_0 arrives last. During the encoding process, the pragmatic encoder generates a three-bit constellation index, $b_5b_4b_3$, for the I symbol coordinate, and simultaneously generates another three-bit constellation index, $b_2b_1b_0$, for the Q symbol coordinate. Note that the length of the input bit sequence must be divisible by five. The bits-to-symbol-constellation map that is applied to the rate 5/6 64-QAM encoder output is shown in Figure 4–23 on page 297.

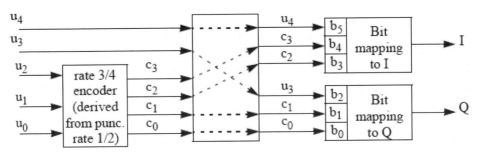

Figure 4–19: TCM encoder for rate 5/6 64 QAM

Figure 4–20 illustrates the rate 3/4 pragmatic TCM encoder for 256-QAM. This encoder uses the rate 1/2 binary convolutional encoder, along with two systematic bits that are passed directly from the encoder input to the encoder output. The sequence of arrival for the $u_2u_1u_0$ input into the encoder is u_2 arrives first, u_1 next, u_0 last. Note that the encoder first generates a four-bit constellation index, $b_7b_6b_5b_4$, which is associated with the I symbol coordinate. Provided another four bit encoder input, it generates a four-bit constellation index, $b_3b_2b_1b_0$, which is associated with the Q symbol coordinate. The I index generation precedes the Q index generation. Note that this encoder should be interpreted as a rate 6/8 encoder, because it generates one eight-bit code symbol per six input bits. For this reason, the length of the input bit sequence should be divisible by six. Figure 4–24 (on page 298) depicts the bits-to-symbol-constellation map that is applied to the rate 3/4 256-QAM encoder output.

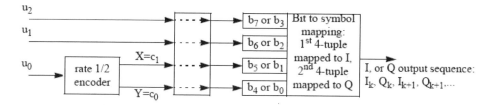

Figure 4–20:   Optional TCM encoder for rate 3/4 256 QAM

Next, Figure 4–21 illustrates the rate 7/8 pragmatic TCM encoder for 256-QAM. This encoder uses a rate 3/4 punctured version of the baseline rate 1/2 binary convolutional encoder, along with two systematic bits that are passed directly from the encoder input to the encoder output. The rate 3/4 punctured code is generated from the baseline rate 1/2 code using an appropriate puncture mask. Puncture samples are sequenced c_3 first, c_2 second, c_1 third, and c_0 last. The sequence of arrival for the $u_6 u_5 u_4 u_3 u_2 u_1 u_0$ input into the encoder (as a whole) is u_6 arrives first, u_5 arrives second, u_4 arrives third, u_3 arrives fourth, u_2 arrives fifth, u_1 arrives next to last, and u_0 arrives last. During the encoding process, the encoder generates a four-bit constellation index, $b_7 b_6 b_5 b_4$, for the I symbol coordinate, and simultaneously generates another four-bit constellation index, $b_3 b_2 b_1 b_0$, for the Q symbol coordinate. The length of the input bit sequence must be divisible by seven. Figure 4–24 depicts the bits-to-symbol-constellation map that is applied to the rate 7/8 256-QAM encoder output.

To make the average transmitted power identical and equal to 1 for all modulation formats, the constellation points are multiplied by an appropriate constant. The QAM symbols are next framed within a message burst, which typically introduces additional framing symbols. The burst symbols are then multiplexed into a duplex frame, which may contain multiple bursts. The I and Q components are injected into pulse-shaping filters, quadrature modulate a certain carrier frequency, and amplified with power control so that the proper output power is transmitted.

The WirelessMAN-SC2 standard envisions the possibility that FEC would not be necessary if ARQ is enabled. In this case, the bits after scrambling are directly mapped to a QAM constellation using appropriate Gray coding. In this case, for BPSK, QPSK, and 16 QAM, the mapping in Figure 4–22 is used, but for 64 QAM and 256-QAM the mappings in Figure 4–25 and Figure 4–26 are used. No FEC operation is mandatory for QPSK; no FEC is optional for other modulation methods.

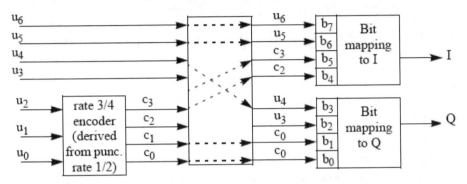

Figure 4–21: Optional rate 7/8 TCM encoder for 256 QAM

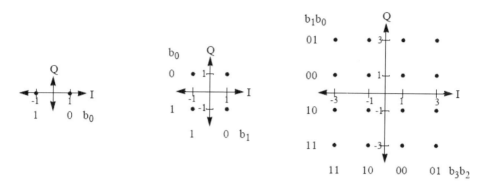

Figure 4-22: Gray maps for BPSK, QPSK, and 16 QAM constellations

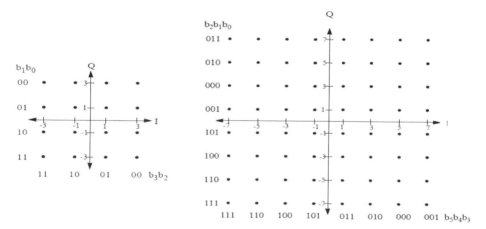

Figure 4-23: Gray-coded symbol maps for 1/2 16 QAM and 64 QAM

Chapter 4: Air Interface for Fixed Broadband Wireless Access Systems

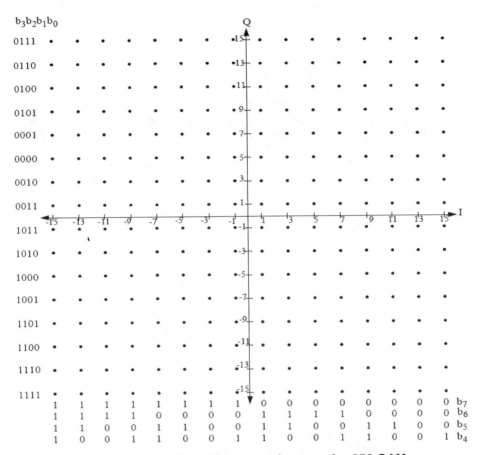

Figure 4–24: Bit-to-symbol mapping map for 256 QAM

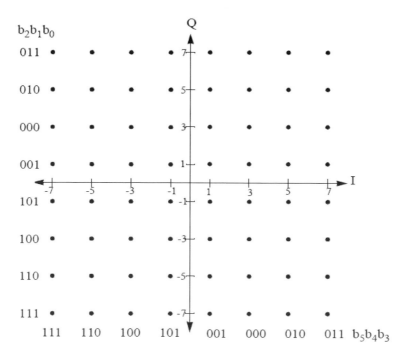

Figure 4–25: Gray coding for 64 QAM when no FEC is used

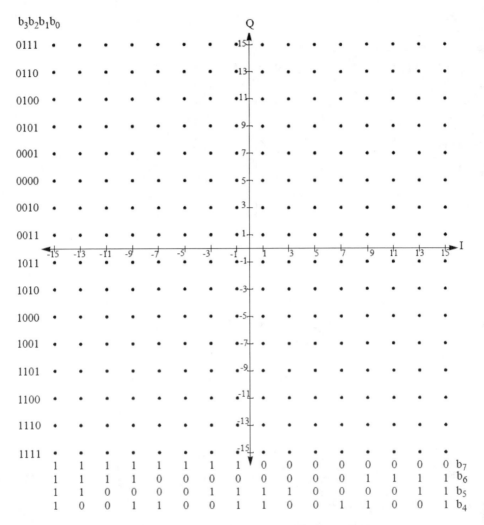

Figure 4–26: Gray coding for 256 QAM when no FEC is used

To achieve higher coding gain, optional turbo product coding (TPC) can be used for all modes. The constituent codes in the turbo coding case are extended Hamming codes or parity check codes. Many soft decoding algorithms can be used for turbo decoding. To match an arbitrary required packet size, BTCs can be shortened by removing symbols from the BTC array. Rows, columns, or parts thereof can be removed until the appropriate size is reached. For transmission, an encoded BTC block shall be read out of the encoded matrix as a serial bit stream, starting with the least significant bit and ending with the most significant bit. This bit stream is sent to a symbol mapper the same as the symbol mappers in the no-FEC case. If not enough encoded bits are available to fill the last symbol of an allocation, sufficient zero-valued bits (unscrambled) will be appended to the end-of-the-serial stream to complete the symbol.

Another option in 802.16 is using convolutional turbo codes. The convolutional turbo code encoder and its constituent encoder are depicted in Figure 4–27 and Figure 4–28, respectively. It uses a double binary circular recursive systematic convolutional code.

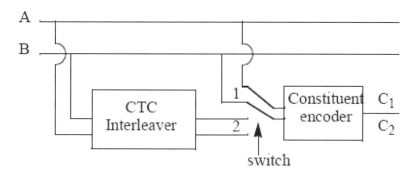

Figure 4–27: A convolutional turbo encoder

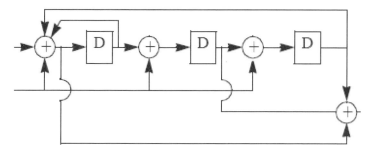

Figure 4–28: A constituent convolutional encoder

During the encoding process, the bits of the data are alternately fed to A and B, starting with the msb of the first byte being fed to A. The encoder is fed by blocks of k bits or N couples ($k = 2N$ bits). For all frame sizes, N is a multiple of 4 and limited between 8 and 256. After initialization, the encoder is fed the sequence in the natural order (position 1) with the incremental address $i = 0,...,N-1$. This first encoding is called C1 encoding. Then the encoder is initialized differently and is fed by the interleaved sequence (switch in position 2) with incremental address $j = 0,...,N-1$. This second encoding is called C2 encoding. Code-rates of 1/2, 2/3, 3/4, 5/6, and 7/8. are achieved by selectively deleting the parity bits (puncturing). The puncturing patterns are identical for both codes C1 and C2.

Both DL and UL data are formatted into bursts that use the framed burst format. The DL supports the most general case of TDM bursts, while the UL supports TDMA bursts. TDMA burst and continuous DL operational modes are subclasses of the TDM burst DL mode. The burst frame consists of a burst preamble and burst payload. Pilot words can be inserted in the payload. A pilot word is a unique word, known to the receiver, and used to help signal processing operations at the receiver. The length in symbols, U, of a unique word (UW) is a burst profile parameter. For best performance, U should be at least as long as the intended channel's delay spread. 802.16 specifies that unique words have lengths at powers of two. These unique words have I and Q components, where

$$I[n] = \cos(\theta[n]) \qquad \text{Eq. 4–6}$$

$$Q[n] = \sin(\theta[n]) \qquad \text{Eq. 4–7}$$

When the length L is 16, 64, or 256, the unique words are derived from Frank–Zadoff sequences, $\theta_{FZ}[n] = (2\pi pq)/\sqrt{L}$. It is clear that when the length is 16, the unique word consists of 4-PSK (equivalent to QPSK) symbols, and when the length is 64 or 256, the unique word consists of 8-PSK and 16-PSK symbols respectively. The sequence length of 64 is mandatory. Length of 16 is also mandatory for symbol rates below 1.25 Msymb/s, and a length of 256 is mandatory for symbol rates above 20 Msymb/s.

Optionally, unique words with lengths of 8, 32, 128, or 512 derived from Chu sequences can also be used, $\theta_{Chu}[n] = \pi n^2/L$. All of these unique words are chosen because they can be easily detected due to their constant amplitude zero autocorrelation (CAZAC) properties [B108].

A downlink burst may contain time division multiplexed messages that are adaptively modulated for the intended message recipients. When a MAC frame control message is to be transmitted within a downlink burst, it will always appear first and must use QPSK. Subsequent messages within the burst shall be sequenced in decreasing order of modulation robustness, beginning with the most robust modulation that is supported at the transmitter. An uplink burst contains a single message and uses a single modulation format within a burst.

Prior to modulation, the I and Q signals are filtered by a square-root raised cosine filter with a roll-off factor of 0.25. Roll-off factors of 0.18 and 0.15 are optional. The transmitted waveform is obtained using quadrature modulation.

802.16 does require power control of the transmitted waveform, but leaves the details of the algorithm to vendors. It also requires channel quality measurements (RSSI and CINR) to be performed.

Antenna diversity is another technique that is likely to find application in BWA systems. To provide another dimension of flexibility, the WirelessMAN-SC2 physical layer supports antenna diversity as well.

With two-way delay transmit diversity, two transmit antennas are used, and the output of the second antenna is delayed with respect to the first. Receive diversity does not require any changes in the physical layer. The receiver equalization is expected to handle the extra delay spread introduced in the system due to the delayed output of the second antenna.

However, when the optional STC transmit diversity scheme [B4] is used, a different framing mechanism is required. The STC transmit diversity scheme logically pairs blocks of data separated by delay spread guard intervals. These paired blocks are jointly processed at both the transmitter and receiver. The technique to be described is particularly amenable to a frequency domain equalizer implementation. Suppose that $s_0[n]$ and $s_1[n]$ are the blocks to be transmitted, each of length F symbols. A two antenna transmitter can transmit these two sequences as follows. The first antenna will transmit $s_0[n]$ and $s_1[n]$, separated by guard intervals (Figure 4–29). The length of these guard intervals is equal to the length of the unique word, which, as has been mentioned, should be equal to the length of the channel impulse response. Instead of guard intervals composed of zeros, it is better to use cyclic prefixes or unique words.

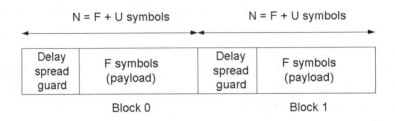

Figure 4-29:   Paired blocks in STC transmit diversity combining

The second antenna will reverse the order in which sequences are transmitted, and will conjugate and time-reverse each sequence. Therefore the first block transmitted by the first antenna will be $-s_1^*[L-n]$ and the second block will be $-s_0^*[L-n]$. At the receiver, frequency-domain equalization can be used to recover the transmitted signal, provided that the channel is known.

A STC burst consists of a preamble, followed by a payload. The preamble consists of several unique words, and the number of such unique words is a burst profile parameter. The payload may consist of multiple pairs of blocks as in Figure 4-29. The preamble structure may also be inserted within a transmission as a group of contiguous pilot words to assist in the channel estimation. In this case, the group of contiguous pilot words is considered external to the paired data blocks. The pilot word repetition interval and the number of unique words comprising one pilot are burst profile parameters.

STC-encoded data and conventionally encoded data will not be time-division multiplexed within the same burst.

WirelessMAN-OFDM PHY

The WirelessMAN-OFDM physical layer is designed for NLOS operation between 2 GHz and 11 GHz and is based on OFDM modulation. It can be used in unlicensed and licensed bands.

The OFDM symbol duration, or the related carrier spacing in frequency, is a major design parameter of an OFDM system. The symbol duration is composed of the FFT interval and of the cyclic prefix (CP). The presence of a cyclic prefix leads to an equivalent loss of SNR, because the cyclic prefix at the receiver is discarded and therefore part of the transmitter energy is wasted. The advantages of having a cyclic prefix are many (multipath immunity, better detection and synchronization performance) and outweigh this energy loss. Note that in the downlink, the BS is not allowed to change the length of the cyclic prefix, because the SS will lose synchronization.

One difference in IEEE 802.16 with respect to IEEE 802.11a is that the OFDM symbol is divided into logical subchannels to support scalability, multiple access, and advanced antenna array-processing capabilities.

An OFDM symbol consists of data carriers, pilot carriers, and null carriers (Figure 4–30). The null carriers, the DC carrier and the guard carriers, are not used. Four parameters describe the OFDM symbol. The number of FFT points is 256, out of which 200 are used, with 55 reserved as guard bins. The DC carrier is also unused. Within the 200 used carriers, 8 carriers are used for pilots. The cyclic prefix can be of lengths 64, 32, 16, or 8 carriers.

Scrambling, FEC, and interleaving are performed prior to modulation. The same randomizer is used as in the physical layer for frequencies between 10 GHz and 66 GHz, described in the previous section. On the downlink, the scrambler is initialized at the start of every frame with 100101010000000. On the uplink, the scrambler is initialized with the vector in Figure 4–31.

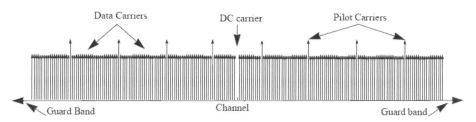

Figure 4–30: An OFDM symbol.

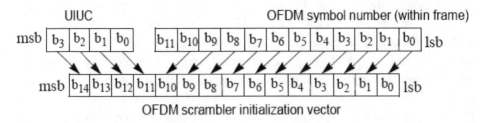

Figure 4–31:   Scrambler initialization on the uplink

A FEC, consisting of the concatenation of a RS (255, 239) outer code over GF (256) and a rate-compatible convolutional inner code, is supported on both the uplink an downlink. Support of block turbo codes with extended Hamming or parity check codes is an option. Convolutional turbo codes are also optional. The block turbo code and the convolutional turbo code are the same as the codes in WirelessMAN-SC2. The Reed–Solomon–Convolutional coding rate 1/2 must always be used when requesting access to the network and in the FCH burst. The encoding is performed by first passing the data in block format through the RS encoder and then passing it through a zero-terminating convolutional encoder. Shortening and puncturing can be performed. After bit interleaving, the data bits are entered serially to the constellation mapper. Gray-mapped QPSK and 16 QAM, as shown in Figure 4–32, are mandatory for this physical layer, while 64 QAM is optional. The constellations shown in Figure 4–32 must be normalized by multiplying the constellation point with the indicated factor c to achieve equal average power. Adaptive modulation and coding must be supported in the downlink per SS. In the uplink, each SS can use different modulation schemes based on the MAC burst configuration messages coming from the BS. After modulation the symbols are mapped onto the carriers. The coding and modulation parameters are summarized in Table 4–5.

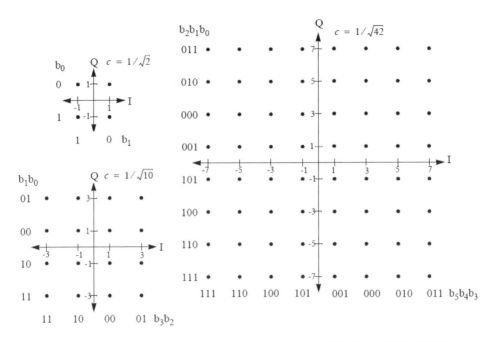

Figure 4–32:   QPSK, 16 QAM, and 64 QAM for WirelessMAN-OFDM

Table 4–5:   Modulation and coding parameters for WirelessMAN-OFDM

Modulation	Uncoded block size (bytes)	Coded block size (bytes)	Overall coding rate	RS code	CC code rate
QPSK	24	48	1/2	(32,24,4)	2/3
QPSK	36	48	3/4	(40,36,2)	5/6
16 QAM	48	96	1/2	(64,48,8)	2/3
16 QAM	72	96	3/4	(80,72,4)	5/6
64 QAM	96	144	2/3	(108,96,6)	Ω
64 QAM	108	144	3/4	(120,108,6)	5/6

Pilots are inserted in the OFDM symbols in the following way. A pseudorandom binary sequence generator of length 11 is used, which is described by the polynomial $1 + x^9 + x^{11}$. The initialization sequences are all ones in the downlink and 10101010101 in the uplink. Starting with the third bit, the output of this random sequence generator, w_k, is used to obtain the pilot tones in the first OFDM symbol following the frame preamble. The pilot tones are modulated using BPSK. On the downlink $c_{-84} = c_{-36} = c_{60} = c_{84} = 1 - 2w_k$ and $c_{-60} = c_{-12} = c_{12} = c_{36} = 1 - 2\overline{w_k}$. On the uplink $c_{-84} = c_{-36} = c_{12} = c_{36} = c_{60} = c_{84} = 1 - 2w_k$ and $c_{-60} = c_{-12} = 1 - 2\overline{w_k}$.

Two preambles are defined—short and long. The short preamble is used in the uplink and consists of 128 samples repeated twice and preceded by a cyclic prefix (CP). The length of the cyclic prefix is the same as the length of the cyclic prefix of the payload-carrying OFDM symbols. The long preamble is used in the downlink. It consists of a CP, followed by 64 samples, repeated four times, followed by a CP and 128 samples, repeated twice. (See Figure 4–33 and Figure 4–34.)

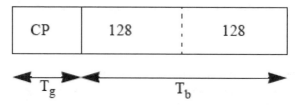

Figure 4–33: Short preamble for uplink data and downlink AAS

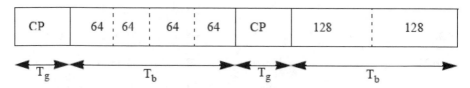

Figure 4–34: Long preamble for downlink and initial ranging

In licensed bands, TDD and FDD can be used. In unlicensed bands, only TDD can be used. The physical layer supports frame-based transmission. One frame consists of a downlink subframe and an uplink subframe. The uplink subframe consists of contention interval for ranging and bandwidth requests, followed by packets from one or more SSs. Every packet consists of a short preamble, followed by an integer number of OFDM symbols. In the downlink, transmissions start with the long preamble, followed by one special OFDM symbol, called FCH symbol. The FCH OFDM symbol conveys the burst parameters and other control information such as the DCD and UCD, and is transmitted using QPSK and rate 1/2 coding. The FCH symbol is followed by one or more bursts, each transmitted with a different burst profile. Each burst contains an integer number of OFDM symbols. A TDD frame is shown in Figure 4–35.

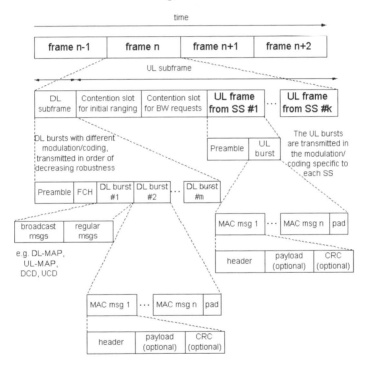

Figure 4–35: An example of an OFDM frame structure with TDD

For FDD, the downlink subframe and the uplink subframe have the same structure as in the TDD case. If FDD is used, instead of time gaps between subframes and frames, the BS must not schedule uplink transmission from a half duplex SS during a downlink allocation.

Mesh networks are an option, and have an optional frame structure, shown in Figure 4–36.

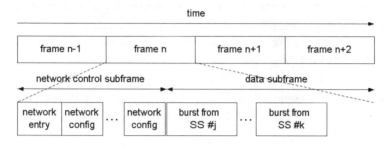

Figure 4–36: Frame structure for mesh networks

A mesh frame consists of a control subframe and a data subframe. The control subframe is used for network creation, maintenance, and coordinated scheduling. All transmissions in the control subframe are done using QPSK and the mandatory coding method. All preambles in the mesh frame are long preambles. Note that distributed scheduling messages can be transmitted in the data subframe as well.

AAS and STC are also supported by this physical layer as options. Power control is required, but not specified and left to vendors.

As an option, STC can be used on the downlink to provide second order spatial transmit diversity. There are two transmit antennas on the BS side and one reception antenna on the SS side. This scheme requires Multiple Input Single Output (MISO) channel estimation. Decoding is very similar to maximum ratio combining. Figure 4–37 shows a block diagram of STC using an OFDM physical layer.

Both antennas transmit two different OFDM data symbols at the same time. The first time antenna 0 transmits s_0, and antenna 1 transmits s_1. The second time antenna 0 transmits $-s_1^*$ and antenna 1 transmits $-s_0^*$. STC is applied independently on each carrier with respect to pilot tone positions. Figure 4–38 serves as illustration, where only one pilot carrier is shown.

Chapter 4: Air Interface for Fixed Broadband Wireless Access Systems

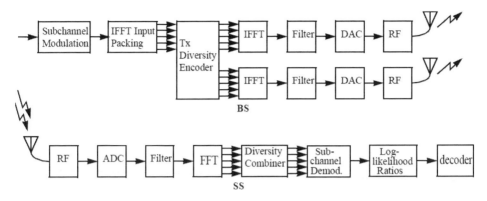

Figure 4–37: A block diagram of a transmitter and receiver constructed according to Alamouti

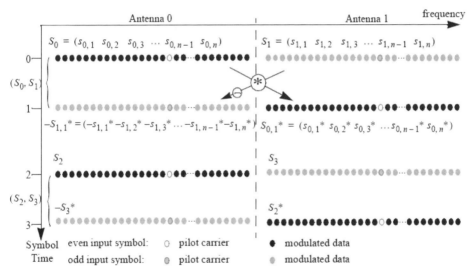

Figure 4–38: STC usage with WirelessMAN-OFDM

Compared with the other wireless standards, in IEEE 802.16 channel quality measurements are much more important. A BS is responsible for making assignments and/or reassignments based on both upstream and downstream channel quality assessments. It can directly determine channel quality on upstream channels from the messages that it receives from SSs. However, to determine the channel quality of its downstream channels, a BS requires channel quality reports from SSs. The BS might also decide on the adaptive modulation type and FEC to be used when communicating with a SS. RSSI and CINR computation is mandatory. RSSI measurements do not require demodulation and can be simply done even at low signal levels. On the other hand, CINR measurements require demodulation. However, CINR is more useful because it provides information on the actual operating condition of the receiver, including interference, noise levels, and signal strength.

Since NLOS subscriber locations may experience Rayleigh or Ricean fading, signal quality (and signal quality measures) may vary with time. For this reason, both mean and standard deviation statistics are defined for the RSSI and CINR measurements. The mean statistics convey the average signal quality level for a channel. The standard deviation statistics are intended to convey the variation in that average value.

The receiver signal processing includes frame detection, fine synchronization, coarse and fine frequency offset estimation and compensation, and channel estimation. This is followed by an orthogonal transformation, phase estimation from the pilot subcarriers, derotation the subcarrier values according to estimated phase, and equalization by dividing each subcarrier value with a complex estimated channel response coefficient. Finally, for each data-carrying subcarrier, find the closest constellation point, compute the Euclidean distance from it and compute the RMS average of all errors in a packet. All of the receiver signal processing operations are similar to IEEE 802.11a. Receiver sensitivity is defined for BER of 10^{-6}. The sensitivity ranges from –65 dBm for 64 QAM with a rate 3/4 code to –91 dBm for QPSK with a rate 1/2 code.

WirelessMAN-OFDMA

WirelessMAN-OFDMA is based on OFDM and designed for NLOS operation in the 2–11 GHz band.

Similar to OFDM, an OFDMA symbol consists of data carriers, pilot carriers, and null carriers. However, OFDM and OFDMA are slightly different. For OFDM, a channel is defined as consisting of all carriers residing within the full signaling bandwidth. In OFDM mode, the basic resource allocation quantum is an OFDM symbol. The amount of data that fits into an OFDM symbol depends on the constellation and the coding method used within this symbol as well as the number of data carriers per symbol. For OFDMA, subchannels are defined as a fraction of the available carriers within the full signaling bandwidth. In OFDMA mode, the basic resource allocation quantum is a subchannel. Furthermore, the carriers forming one subchannel do not have to be adjacent. For all FFT sizes, each OFDMA symbol contains an integer number of subchannels, both on downlink and on uplink. The amount of data that fits into a subchannel depends on the constellation and the coding method used within this subchannel as well as the number of data carriers per subchannel.

Therefore, OFDMA is a multiple access method simultaneously in the time and frequency domains. In OFDMA one slot, also called data region, has two dimensions—time and subchannel. The time dimension denotes consecutive OFDM symbols in the time domain, and the subchannel dimension denotes blocks within the OFDM symbols. Figure 4–39 shows three OFDMA subchannels.

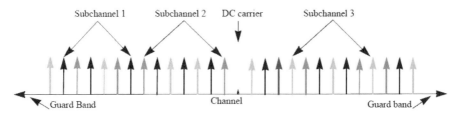

Figure 4–39: An example of three OFDMA subchannels

All used carriers are partitioned into fixed-location pilots, variable-location pilots, and data. In the downlink, the variable-location pilots shift their location every symbol repeating every four symbols. After mapping the pilots, the rest of the used carriers are the data subchannels. Because the variable location pilots change location in each symbol, repeating every fourth symbol, the locations of the carriers in the data subchannels must change also. To allocate the data subchannels, the remaining carriers are partitioned into groups of contiguous carriers. Each subchannel consists of one carrier from each of these groups. The number of groups is therefore equal to the number of carriers per subchannel. The number of the carriers in a group is equal to the number of subchannels.

Regarding the subcarrier allocations in OFDMA, there is a subtle difference between uplink and downlink. This difference is that, in the downlink, the pilot tones are allocated first; what remains are subchannels that are used exclusively for data. In the uplink, however, the set of used carriers is first partitioned into subchannels, and then the pilot carriers are allocated from within each subchannel. Thus, in the downlink, there is one set of common pilot carriers, but in the uplink, each subchannel contains its own set of pilot carriers. This is necessary because, in OFDMA, the BS downlink is broadcast to all SS, but in the uplink, each subchannel may be transmitted from a different SS.

In OFDMA, the FFT size is 2048, out of which 1702 are used in the downlink, and 1696 are used in the uplink. On both the downlink and the uplink there are 32 subchannels, each with 48 data carriers. The cyclic prefix is 1/32, 1/16, 1/8, and 1/4 of the FFT length, or alternatively 64, 128, 256, or 512 samples. In the downlink, there are 142 variable-location pilots, 32 fixed-location pilots, eight of which coincide with the variable location pilots, so the total number of pilots is 166. In the uplink within each subchannel, there are 48 data carriers, one constant-location pilot carrier at index 26, and four variable-location pilot carriers. The variable-location pilots change with each symbol, repeating every 13 symbols. The variable-location pilot locations will never coincide with the constant-location pilot at index 26. The remaining 48 carriers are data carriers.

The signal processing operations are as follows. Each FEC block spans one OFDMA subchannel in the subchannel axis and three OFDM symbols in the time axis. The FEC blocks are mapped in such a way that the lowest numbered FEC block occupies the lowest numbered subchannel in the lowest numbered OFDM symbol. When the edge of a slot is reached, the mapping continues from the lowest-numbered OFDMA subchannel in the next OFDM symbol.

In licensed bands, the duplexing can be TDD or FDD, and in unlicensed bands it can be only TDD.

In PMP TDD operation (Figure 4-40), each frame consists of downlink and uplink, separated by transition gaps. Each burst transmission consists of multiples of three (or six, if STC used) OFDMA symbols. On the downlink subframe, first a FCH is transmitted in the most robust burst profile. The FCH is a data structure that contains the DL-MAP message and can additionally include the UL-MAP, DCD or UCD messages. The DL-MAP message is preceded by a DL frame prefix, which has information where the FCH message is and how it can be demodulated. Since different modulations and coding rates can be used, the transitions between modulations and coding can take place not only on OFDMA symbol boundaries in time domain, but on subchannels within an OFDM symbol in frequency domain. A BS supporting the AAS option will allocate (at the end of the uplink frame) an initial ranging slot for AAS SS that have to initially alert the BS for their presence.

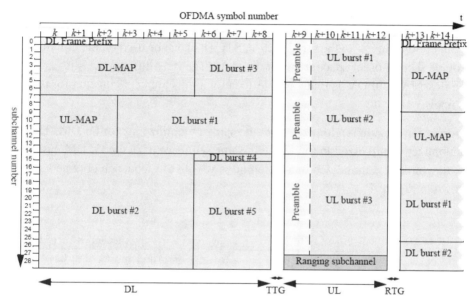

Figure 4–40: An example of a TDD OFDMA frame

When used with the WirelessMAN-OFDMA physical layer, the MAC will define a single ranging channel. There is one ranging channel, composed of one or more pairs of adjacent subchannels, where the index of the lowest-numbered subchannel is even. The indices of the subchannels that compose the ranging channel are specified in the UL-MAP message. Users are allowed to collide on this ranging channel. To effect a ranging transmission, each user randomly chooses one ranging code from a bank of specified binary codes. These codes are obtained from a simple random binary sequence generator, described by the polynomial $1 + x^1 + x^4 + x^7 + x^{15}$ and initialized by the sequence 100101010000000. The first code is formed by the first 106 bits in this sequence;

the next is formed by taking the 107–212 bits; etc. There are 48 codes in all, divided into three usage groups—codes for initial ranging, periodic ranging, and bandwidth requests. The BS dynamically assigns codes to each usage group. These codes are then BPSK modulated onto the carriers in the ranging channel, one bit per carrier. This process is used for initial ranging, periodic ranging, and bandwidth requests. Initial ranging is used by SSs that want to synchronize for the first time. Initial ranging is performed during two OFDM symbols, with the same ranging code transmitted during the two symbols. Periodic ranging and bandwidth requests use one OFDM symbol.

As an option, similar to the OFDM case, STC [B4] can be used on the downlink to provide second-order space transmit diversity. The only difference with WirelessMAN-OFDM is that variable-location pilots are kept constant for two symbols.

Information bits are randomized with the same randomizer as for OFDM. This randomizer is initialized in the DL with same sequence as in the OFDM case. In the downlink it is initialized in a different way, with the sequence obtained as shown in Figure 4–41.

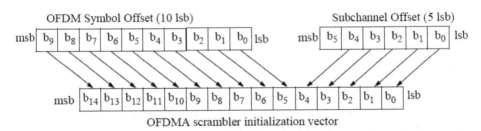

Figure 4–41: Scrambler initialization vector for OFDMA

The FEC mechanism and data modulation formats are similar to those used in the WirelessMAN-OFDM physical layer. Pilot modulation is also similar. The difference in pilot modulation is that each pilot is transmitted with a 2.5 dB boost over the average power of each data tone. This boost is achieved by using $c_k = 8/3 \left(\frac{1}{2} - w_k \right)$, instead of $c_k = 2 \left(\frac{1}{2} - w_k \right)$, as in BPSK.

The OFDMA receiver sensitivity is the same as in the OFDM case.

In the licensed bands, channelization is performed according to the regulatory requirements. The minimum width of one channel is 1.25 MHz. In the middle and upper UNII bands, the same channelization scheme as for IEEE 802.11a is used, with the difference that channels of width 10-MHz can also be used. In unlicensed bands, mesh networks can be used in the same fashion as for WirelessMAN-OFDM.

The same types of channel quality measurements are required (RSSI and CINR), as in the OFDM case.

COEXISTENCE

Radio waves propagate in unbounded medium. Interference among different wireless systems operating in the same frequency band is inevitable. Therefore, it is better if measures are taken to minimize the interference. As discussed in the previous chapter, coexistence is a new issue for wireless communications. Compared with IEEE 802.11 and IEEE 802.15.1, 802.15.3, and 802.15.4, the resolution of coexistence issues is more important for IEEE 802.16 because of the usage scenarios and business model of IEEE 802.16 service providers.

What are the interference mechanisms that can reduce the performance of fixed BWA systems? First, although intrasystem interference is often a significant source of performance degradation, it is not considered. The reason is that intrasystem interference is under the control of the operator. Thus, only intersystem interference mechanisms, where interoperator coordination is appropriate, are considered here. Several coexistence problems should be discussed.

First, there are coexistence issues with other (non-IEEE 802.16) systems, mainly in the unlicensed bands. In the 5-GHz region, satellite services and radar services are two primary users of the bands. The RF center frequency of these satellites is 5.305 GHz; they use antennas with very high gain, and their peak radiated power can be significant. Other devices that also operate in these frequency bands are road transport and traffic telemetric (RTTT) devices [B128]. They are allocated in the 5795–5805 MHz band, with an extension band of 5805–5815 MHz, which in the United States may be used on a national basis at multilane road junctions. These devices are split into the roadside unit (RSU) and the onboard unit (OBU).

Second, there are coexistence issues also among IEEE 802.16 operators, both in the licensed and unlicensed bands. IEEE 802.16 also provides recommendations directed at equipment manufacturers and different recommendations aimed at companies providing services using this technology.

Looking at the sources of interference will be helpful here. Figure 4–42 shows the main sources of interference to a fixed BWA base station, with a sectoral-coverage antenna.

The victim BS is shown as a black triangle on the left, with its radiation pattern represented as ellipses. The desired SS transmitter is shown on lower right of figure. The letters in Figure 4–42 illustrate several cases of interference to a base station.

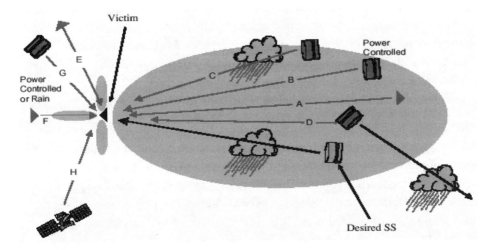

Figure 4–42: Interference sources

Case A is BS-to-BS interference in which each BS antenna is in the main beam of one or more other BSs. This can be a significant problem, because BS antennas tend to be elevated, with a high probability of line-of-sight path to each other. BS-to-BS interference can be reduced by ensuring that there is no BS transmission on frequencies being used for reception at other BSs.

Case B is SS-to-BS interference in which each antenna is in the main beam of the other. As SS antenna gain is much higher than the BS antenna gain, this might appear to be the worst possible case. However, fixed BWA PMP systems can safely be assumed to employ uplink adaptive power control at subscriber stations. This power control equalizes the received signal strength arriving at a BS from near and far SSs on adjacent channels. Note that active control of downlink power from BS transmitters is usually not employed. The net effect is that Case B may not be more severe than Case A. In addition, the narrow beamwidth of a SS antenna ensures that Case B is much less common an occurrence than Case A. However, band planning cannot eliminate Case B interference.

Case C is just like Case B, but the interfering signal is subject to rain attenuation. Therefore it is less of a problem than Case B.

Case D is also like Case B, but the interference is from a SS's sidelobe. However, because this SS sees rain in its main transmission, it does not turn down its power. Case D can be more serious than Case B if the interference does not come from a BWA system, but from a satellite communication system.

Case E is similar to Case A, it is also BS-to-BS interference, but the interference comes from another BS's sidelobe or backlobe. Case E is less serious than Case A.

Case F covers BS-to-BS backlobe-to-backlobe or sidelobe-to-sidelobe. The low gains involved here ensure that this is a problem only for co-deployment of systems on the same rooftop. Like all sources of BS-to-BS interference, this can be virtually eliminated in FDD via a coordinated band plan.

Case G is interference from an SS to the BS's sidelobe or backlobe. It is less severe than Case B and is not considered further.

Finally, **Case H** covers interference from a satellite downlink or stratospheric downlink. This case is not considered in the Recommended Practice.

With the previous simplifying assumptions, the dominant sources of interference that require detailed modeling are shown in Figure 4–43. Case A will tend to dominate unless there is a harmonized band plan for the use of FDD. Case B is always a concern. Case D is less significant than Case B if the source is another BWA system, but could be significant if it comes from a transmitter with high output power. Case F is a concern only for co-sited BSs and can be largely mitigated by the use of a harmonized band plan with FDD.

Figure 4-44 shows the main sources of interference to a subscriber station having a narrow beamwidth antenna. The victim subscriber station is shown along with its radiation pattern (ellipses). The BS and several interferers are also shown. The victim SS cases are fundamentally different from the victim BS cases because the antenna pattern is very narrow. The letters in Figure 4-44 illustrate several cases of interference to a subscriber station:

Case A is SS-to-SS interference, where the beams are collinear. This case rarely happens in practice.

Case B covers BS-to-SS interference.

Case C covers the case of a narrow-beam transmitter (fixed BWA or point-to-point) or satellite uplink at full power, due to rain in its path, but radiating from its sidelobe towards the victim. This case is more likely to occur than Case A because it could occur with any orientation of the interferer.

Case D covers BS-to-SS interference picked up by a sidelobe or backlobe of the victim. This case could be common because BSs radiate over wide areas, and this case could occur for any orientation of the victim.

Case E covers SS-to-SS interference picked up by a sidelobe or backlobe of the victim station. Similar to the reasoning in the victim BS Cases B and C, the worst case can be assumed to be clear air in the backlobe with the interferer having turned its power down.

The source of interference in **Case F** is a satellite downlink. Case F is not considered in the Recommended Practice.

Chapter 4: Air Interface for Fixed Broadband Wireless Access Systems

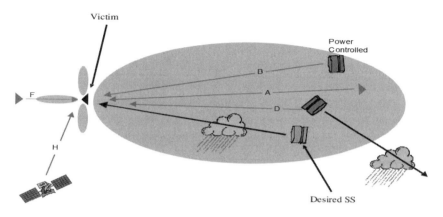

Figure 4-43: Simplified model of the interference to a BS

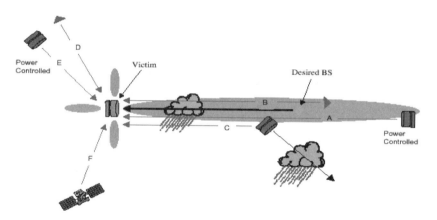

Figure 4-44: Interference to SS

How is the interference calculated? A number of techniques can be used, such as worst-case analysis, Monte Carlo simulations, or the by calculating the interference area. The most appropriate method depends on the interference mechanism. Worst-case analysis is more appropriate when there is a single dominant source of interference. An example is the interference from a single dominant BS into the victim BS of an adjacent system. There are many cases where a simple worst-case analysis is of limited use. Where there are many possible interference paths, the worst case could be very severe but may also be very improbable. An example is the interference between subscriber stations of different operators in the same geographical area. Monte Carlo simulations provide a means of assessing the probability of occurrence of a range of interference levels at victim stations. The recommended geographical or frequency spacing is then a compromise in which an acceptably small proportion of cases suffer interference above the recommended limit. In some scenarios, it can be shown that specific parts of the coverage area will suffer high levels of interference while other areas are not affected. The interference area (IA) is the proportion of the sector coverage area where interference is above the target threshold. This is equivalent to the probability that a randomly positioned station (within the nominal coverage area) will experience interference above the threshold. In several scenarios, the interference area value is a small percentage and the locations are predictable. Although high levels of interference do occur, they are sufficiently localized to be acceptable.

The two types of interference—co-channel and adjacent channel— are shown in Figure 4–45. Note that the channel bandwidth of the co-channel interferer may be wider or narrower than the desired signal. In the case of a wider co-channel interferer (as shown), only a portion of its power will fall within the receive filter bandwidth. In this case, the interference can be estimated by calculating the power arriving at the receive antenna and then multiplying by a factor equal to the ratio of the filter's bandwidth to the interferer's bandwidth. An out-of-channel interferer is also shown. Here a portion of the interferer's spectral sidelobes or transmitter output noise floor falls co-channel to the desired signal; i.e., within the receiver filter's passband. This can be treated as co-channel interference. It cannot be removed at the receiver; its level is determined at the interfering transmitter. By characterizing the power spectral density of sidelobes and output noise floor with respect to the main lobe of a signal, this form of interference can be approximately computed in a manner similar to the co-channel interference calculation, with an additional attenuation factor due to the suppression of this spectral energy with respect to the main lobe of the interfering signal.

There are two types of unwanted emissions. First, out-of-band emissions are within 200% of the bandwidth from the edge of the authorized emissions. Second, there are spurious emissions, which are beyond this point. The spectral density of unwanted emissions at the input to the antenna port should be attenuated significantly. Spurious emissions should not exceed an absolute level of −70 dBW/MHz in any 1 MHz band.

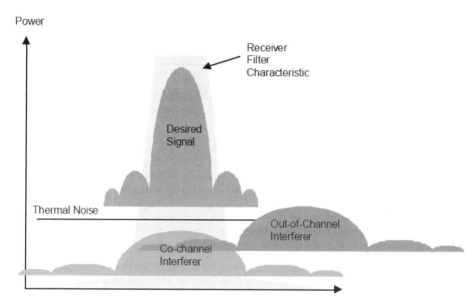

Figure 4–45: An illustration of co-channel and out-of-channel interferer

In the unlicensed portion of the 2–11 GHz band, coexistence issues are especially important, because any wireless system can operate in this portion of the band. Two categories of coexistence mechanisms are considered—methods that fall within the scope of the WirelessMAN standard and those that fall outside that scope. Within the scope of the WirelessMAN standard, two methods are identified—DFS and TPC. For unlicensed use, DFS and TPC are required both by the standard and by regulatory bodies in the United States and Europe. Outside the scope of the WirelessMAN standard, two methods are identified—antenna directivity and antenna polarization. Here is a summary of the four methods.

1) *DFS*. As frequency planning is not practical in licensed-exempt bands, DFS can be used to avoid assigning a channel to a channel occupied by another system. DFS is generally based on comparison of a C/I threshold against idle time RSSI measurements. DFS is predominantly effective to combat interference from and to ground based systems, such as WLANS, RTTT, radar

and other wireless systems. It is generally ineffective to combat interference from and to airborne systems, such as airborne radars and satellites.

2) *TPC.* This is a technique to adaptively adjust the transmit power of a transmitter to maintain the received signal level within some desired range. With power control, the transmitter EIRP is reduced according to the link margin. Shorter link ranges, hence, result in lower transmitted power levels. For PMP systems, the average EIRP will typically be several decibels below the legal limit. For mesh systems, the EIRP values decrease rapidly as customer deployment density increases. As power control is also influenced by C/I levels, the use of adaptive power control with DFS, where possible, tends to result in the most effective interference mitigation.

3) *Antenna directivity.* Antenna directivity, in horizontal but especially in vertical direction can significantly reduce a BWA's interference potential and resilience. Vertical directivity especially reduces the interference caused to satellite systems, which are designated primary users of part of the addressed bands. It also can significantly help reduce interference to and from indoor WLAN systems. Horizontal directivity significantly reduces the probability of interference to other systems (assuming interference is mainly caused in the main lobe), but tends to increase the severity of the interference, as the energy in the main lobe is generally higher.

4) *Antenna polarization.* Antenna cross-polarization in the 5GHz band can achieve an isolation of up to 15 dB in LOS, but the isolation reduces significantly in near-LOS and NLOS environments. Most deployments use both horizontal and vertical polarization to maximize spectral reuse. Polarization hence has the potential to provide some isolation between differently polarized systems, especially in LOS. However, given the operational needs and implementation of most systems in the targeted spectrum, the effectiveness will be mostly marginal.

While in the unlicensed bands there will be interference from other wireless systems, in the licensed bands coexistence issues among multiple operators will dominate. It is noted that in general some service providers may deploy systems not conforming to IEEE 802.16. The coexistence problem in this case becomes very complex and is not considered. Because interference among operators can be a contentious issue, IEEE 802.16 produced a Recommended Practice called IEEE Std 802.16.2. It should be noted that the coexistence issue is so complex in general that it cannot be addressed in a standard. These recommendations can be considered as a system model. The input to the model is the system architecture, transmitted power, spectral masks and antenna patterns, and receiver parameters, such as noise floor degradation and blocking performance. The outputs of the model are recommendations for the deployment of BWA systems. These recommendations include band plans, separation distances, and power spectral flux density limits. The information is therefore valuable as a first step in planning the deployment of systems. In the model calculations, the geographical or frequency spacing between systems is being varied until the interference is below an acceptable threshold.

There are two main interference scenarios. First, there is a co-channel scenario, in which two operators are in either adjacent territories or territories within radio line of sight of each other and have the same spectrum allocation. Second, there is an adjacent channel scenario in which the licensed territories of two operators overlap and they are assigned adjacent frequency bands. Coexistence issues may arise simultaneously from both scenarios as well as from these scenarios involving multiple operators. These, however, are very complex, and are not considered. First, the general recommendations, applicable in all cases, are considered. Then, the recommendations for the co-channel and adjacent channel cases are considered separately.

The general recommendations are:

1) Each operator should accept a 1-dB degradation [the difference in dB between C/N and $C/(N + I)$] due to cumulative interference from other operators. Operators should include 1-dB margin even if the first to deploy in a region. Thus interference that produces a degradation of 1 dB to the system's C/N is considered acceptable. The logic here is that there is no such thing as interference-free wireless transmission. Therefore, if the interference is 6 dB below the noise, it can be said that systems coexist. For the noise floor to increase by 1 dB, the interference power level must be 6 dB below the receiver's thermal noise floor. According to the link budget calculations for wireless system (previously presented), the level of the received signal will decrease depending on the distance and the atmospheric conditions (rain, etc.). The receiver will receive the signal only when the signal-to-noise ratio exceeds a certain threshold. A way to account for interference is to determine

$C/(N + I)$, the ratio of carrier level to the sum of noise and interference. Receiver sensitivity determines the minimum detectable signal and is a key factor in any link design. However, as the level of receiver noise floor increases, the sensitivity degrades. This, in turn, causes reduction in cell coverage, degradation in link availability, and loss of revenues. The factors contributing to the increase in noise power divide into two groups—internal and external. The internal factors include the thermal noise generated by various components within the receiver, intermodulation noise, and intranetwork co- and adjacent-channel interference. The external factor is internetwork interference. The amount of degradation in receiver sensitivity is directly proportional to the total noise power added to the thermal noise, I, consisting of intra- and internetwork components. In order to reduce the internetwork contribution, it is recommended that the effect of any fixed BWA network on any other coexisting BWA network should not degrade the receiver sensitivity of that fixed BWA network by more than 1 dB.

For example, consider a receiver with 6 dB noise figure. The receiver thermal noise is –138 dBW/MHz. Interference of –138 dBW/MHz would double the total noise, or degrade the link budget by 3 dB. Interference of –144 dBW/MHz, 6 dB below the receiver thermal noise, would increase the total noise by 1 dB to –137 dBW/MHz, degrading the link budget by 1 dB.

2) Each operator should collaborate in good faith with other known operators before initial deployment and before every relevant system modification. In general, analyses to evaluate the potential for interference and any possible mitigation solution should be performed before system implementation. Coordination with adjacent operators could significantly lower the potential for interference. Best results may be obtained if full cooperation and common deployment planning is achieved. It is legitimate to consider the capital investment an incumbent operator has already made and the additional investment required by an incumbent operator for a change enhancing the ability to coexist versus the capital investment costs that the new operator would incur.

3) The degree of coexistence between systems depends on the emission levels of the various transmitters. Thus it is important to recommend an upper limit on transmitted power or, more accurately, a limit for the equivalent isotropically radiated power (EIRP). Because point-to-multipoint systems span very broad

frequency bands and use many different channel bandwidths, a better measure of EIRP for coexistence purposes is in terms of *power spectral density* (psd) expressed in dBW/MHz rather than simply power in dBW. Note that the regulatory limits are significantly higher (e.g., 15 dB) than supported by most currently available equipment. If systems are deployed using the maximum regulatory limits, they should receive a detailed interference assessment unless they are deployed in isolated locations, remote from adjacent operators. In the coexistence simulations, reasonable cell sizes and link budgets are considered. Typical parameters used for the BS and in coexistence simulations for the recommended practice are as follows—Tx Power: +24 dBm (–6 dBW), SS Antenna Gain: +34 dBi, BS Antenna Gain: +19 dBi, and Carrier Bandwidth: 28 MHz (+14.47 dB-MHz). (These design parameters apply to frequency range 2 (23.5–43.5 GHz.) According to 802.16.2, a BS should not produce an EIRP power spectral density exceeding +14 dBW/MHz. However, it is strongly recommended that a maximum EIRP power spectral density of 0 dBW/MHz be used. A SS should not produce an EIRP spectral density exceeding +30 dBW/MHz. However, it is strongly recommended that a maximum EIRP power spectral density of +15 dBW/MHz. (For the specific sub-band 25.25–25.75 GHz, a subscriber station should not exceed +24 dBW in any 1 MHz bandwidth [ITU-R F.1509]. Repeaters with directional antennas should not produce an EIRP spectral density exceeding +30 dBW/MHz, but a value of +15 dBW/MHz is recommended. Repeaters employing omnidirectional or sectored antennas should not produce an EIRP spectral density exceeding +14 dBW/MHz, with a recommended value of 0 dBW/MHz. Furthermore, a SS should employ uplink power control with at least 15 dB of dynamic range.

4) The antennas play an important role in the coexistence. Independently of polarization, the operator needs to consider the antenna radiation pattern in the azimuth and elevation planes relative to the required coverage footprint. Two linear polarization orientations, horizontal and vertical, are recommended. The required polarization purity is captured in the specification of antenna cross-polar discrimination (XPD). The XPD of an antenna for a given direction is the difference in dB between the peak co-polarized gain of the antenna and the cross-polarized gain of the antenna in the given direction. Cross-polarization can be effective in mitigating interference between adjacent systems. A typical cross-polarization isolation of 25–30 dB can be

achieved with most antennas today. This is sufficient to counter co-channel interference for QPSK and 16-QAM modulation schemes. Also, the Radiation Pattern Envelopes (RPEs) are independent of polarization. The performance of BS antennas is here divided into two electrical classes. Class 1 represents the minimum recommended performance. Class 2 antennas have enhanced RPEs and represent more favorable coexistence performance. Class 2 antennas are recommended for operation in environments in which interference levels could be potentially significant. Fixed BWA systems employ SS antennas that are highly directional, narrow-beam antennas. Although it is not as important for coexistence as the BS RPE, the RPE of the SS antenna is a factor in determining intersystem interference.

The performance of SS antennas is divided into three electrical classes. Class 1 is defined with moderate sidelobe characteristics and represents the minimum recommended performance. Class 2 and Class 3 antennas have enhanced RPEs and represent increasingly favorable coexistence performance. Most of the coexistence problems are the result of main-beam interference. The side lobe levels of the base station antennas are of a significant but secondary influence. The side lobe levels of the subscriber antenna are of tertiary importance.

Furthermore, mechanical parameters of the antenna also can affect coexistence. The mechanical characteristics that are considered are wind and ice loading, water ingress, temperature and humidity, and vibration. For deployments in hurricane-prone areas, more robust antenna systems may be required.

Factors related to the antennas can also be efficient mitigation techniques. These are antenna-to-antenna isolation, orientation, tilting, directivity, and polarization. Out of these changing the antenna orientation is especially effective in the case of interference arising from main-beam coupling. Near the service area boundaries, coexistence is improved with a high-performance antenna with high directivity as opposed to a broader range sectorized antenna or omnidirectional antenna.

5) It is recommended that the subscriber and base station equipment have the ability to detect and react to failures, either software or hardware, in a manner

Chapter 4: Air Interface for Fixed Broadband Wireless Access Systems

to prevent unwanted emissions and interference. For example, in the absence of a correctly received downlink signal, the SS transmitter should be disabled.

These five recommendations are valid in all cases. Consider specifically the following co-channel and adjacent channel scenarios.

1) *The co-channel case.* The most problematic interference occurs between base stations. Careful deployment of fixed BWA systems is important in minimizing interference problems. Coordination is recommended between licensed service areas where both systems are operating co-channel (i.e., over the same fixed BWA frequencies) and where the service areas are in close proximity, (e.g., the shortest distance between the respective service boundaries is less than 60 km). The operators are encouraged to arrive at mutually acceptable sharing agreements that would allow for the provision of service by each licensee within its service area to the maximum extent possible. Under the circumstances where a sharing agreement between operators does not exist or has not been concluded, and where service areas are in close proximity, a coordination process should be employed. As described previously, distance is suggested as the first trigger mechanism for coordination between adjacent licensed operators. If the boundaries of two service areas are within 60 km of each other, then the coordination process is recommended. The number 60 km is obtained by calculating the radio horizon by $R = 4.12(\sqrt{h_1} + \sqrt{h_2})$, where h_1 and h_2 are in the heights of the antennas above clutter in meters, and R is in kilometers. Assuming that the BS antennas are 55 m above the radio clutter, the radio horizon will be 60 km. If the base stations are separated by a distance larger than the radio horizon, coexistence problems among them will not arise. Other forms of interference are still possible (for example, SS to BS), but they will be generally less severe. It is suggested that service area demarcation is used. IEEE 802.16 also recommends that government regulators specify trigger values of for each frequency band. If the trigger values are exceeded, the operator should try to modify the deployment or should coordinate with the affected operator. The coordination trigger values of –114 (dBW/m)/MHz (24, 26, and 28 GHz bands) and –111 (dBW/m 2)/ MHz (38 and 42 GHz bands) are recommended. These "triggers" should be applied before deployment and before each relevant system modification. If these trigger values are exceeded, there are several possible solutions. Increasing the distance between the interfering

transmitter and receive would obviously result in decreased interference. If the distance cannot be increased, the same effect can be achieved by lowering the transmitter power. As a mitigation technique in the co-channel case, frequency exclusion is a possible, albeit very undesirable, approach for avoiding interference. This practice involves dividing or segregating the spectrum so that neighboring licensees operate in exclusive frequencies, thus avoiding any possibility for interference. Instead of frequency exclusion, by retaining spare frequencies to use only when interference is detected, some potential co-channel and adjacent channel problems can be eliminated.

2) *Adjacent channel case in which the served territories overlap.* When operators are in the same area but on adjacent frequency bands, analysis and simulation indicate that deployment may require a guard frequency. In most co-polarized cases, the guard frequency should be equal to one equivalent channel of the widest bandwidth system. Where administrations do not set aside guard channels, the affected operators would need to reach agreement on how the guard channel is apportioned between them. This situation has one interesting special case—when multiple operators share an antenna site by using harmonized transmissions. For FDD systems, this implies that each operator's base station transmits in the same frequency sub-block (typically on a different channel) and that their terminals transmit in the corresponding paired sub-block. For TDD systems, harmonization implies frame, slot, and uplink/downlink synchronization. It is clear that planning is required for co-sited antennas of BSs. When tackling coexistence between FDD systems operating in adjacent frequency blocks in the same or overlapping areas with defined uplink and downlink frequency bands, co-siting of base station transmitters actually facilitates coexistence.

Alternative schemes might be available in the future. For example, technologies such as adaptive arrays or beam-steering antennas can focus a narrow beam towards individual users throughout the service area in real-time to avoid or minimize coupling with interfering signals. Beam-shaping arrays, which create a null in the main beam towards the interfering source, represent another possible approach towards addressing interference. Using more robust modulation and enhanced signal processing techniques can help in deployment scenarios where the potential for interference is high.

BWA BUSINESS AND TECHNOLOGY TRENDS

IEEE 802.16 is currently interested in the following technologies. First, all physical layers at present work above 2 GHz. However, it is possible and economically justified to have operation below 2 GHz. The MAC of IEEE 802.16 may be modified to support point-to-point operation [B133], and not only point-to-multipoint. Another initiative within IEEE 802.16 is to support mobility. Task Group 802.16e was established to support this technology [B30], [B97], [B99], [B131]. The solution that will be developed by 802.16e will likely be devoting one part of the superframe for fixed stations and one part for mobile stations. The further technical development of IEEE 802.16 will likely take place in these areas.[11]

There will be one billion Internet users in the year of 2005, and even more in future years. A relatively small, but growing percentage of them are accessing the Internet using broadband technology. Therefore, the potential market for WMAN is significant. To promote interoperability among IEEE 802.16 devices and to promote WMAN technology based on IEEE 802.16, several companies, including Intel, Fujitsu®, and Nokia, have formed a trade organization called WiMAX Forum. This trade organization will certify the compliance of IEEE 802.16 equipment and will promote it in the marketplace. Some analysts expect the first IEEE 802.16 products to be available in 2004–2005 and expect the total market to reach $1 billion by 2008. It remains to be seen whether these market predictions will be correct. IEEE 802.16 is very suitable for developing countries that lack telephone and cable infrastructure. In developed countries, the market opportunity in rural and suburban areas lacking DSL and cable is also significant. Even in urban areas, where DSL and cable services are available, many users will choose wireless because of its convenience. For example, while DSL works at distances up to 18 000 feet from a telephone company's central office, IEEE 802.16 works at distances up to 30 km from a base station.

One reason the broadband wireless Internet access market in the United States and other countries has remained small has been the lack of a standard, especially during the years of high growth in the telecommunications industry.

---

11. For the latest information on the 802.16 Working Group and ongoing projects, see http://grouper.ieee.org/groups/802/16/.

Chapter 4: Air Interface for Fixed Broadband Wireless Access Systems

BWA business is different from the WLAN and WPAN businesses in one important aspect. The use of WPAN and WLAN devices does not require monthly service fees. As a result, service providers have been generally struggling to devise ways to raise revenue from WLAN and WPAN service. On the other hand, because BWA devices are used for Internet access, their use will certainly involve a monthly service fee. In addition, BWA can enable other revenue-generating services, such as interactive games, video on demand, etc. To make BWA systems even more competitive for the "first mile" connection to services, IEEE 802.16 provides maximum flexibility and relatively low "up-front" costs.

Overall, it seems plausible that, one way or another, wireless Internet access will succeed in the marketplace. At present in the U.S., UK, Sweden, and other countries, wireless broadband service is available in isolated areas. Examples in the U.S. are small service providers such as Mesh Networks Inc., in Florida, BroadBand Solutions in Utah, and other companies. These networks are not based on IEEE 802.16, but they support mobility, providing coverage at speeds of about 40–50 mph.

One question that remains unanswered is whether having licensed spectrum and unlicensed spectrum versions of the same service makes business sense. Some telecommunication service providers that bought spectrum at significant cost soon afterwards found themselves in an untenable business environment and ceased operation altogether.

In general, IEEE 802.16 makes the competition and cooperation with 3G and 4G cellular technologies much more interesting. Communications anytime, anywhere clearly requires wireless networks to seamlessly connect with other wireless networks. Flarion Technologies of Bedminster, NJ, and SOMA Networks™ of San Francisco, CA, are some of the companies that offer so-called 4G technologies, which integrate cellular communications and wireless data.

IEEE 802.16's major weakness is that it does not support mobility. While originally mobility was not required, recently demand has been increasing for mobile broadband wireless access. Within IEEE 802. there are two efforts to provide mobile BWA systems. First, IEEE 802.16 has created a task group to explore how to modify IEEE 802.16 so that it supports mobility. Also, a separate working group was established, IEEE 802.20, with the goal to develop an air interface for a technology achieving data rate of 1 Mb/s at speeds up to

250 km/hour at cell ranges of up to 15 km. The typical application is Internet access in trains and other fast-moving vehicles. IEEE 802.20 will work in licensed bands below 3.5 GHz. The modulation technologies will achieve spectral efficiencies of more than 1 bit/second/Hz/cell, which is more than double than that of today systems. IEEE 802.20 will support voice, video, and data services. VOIP, online gaming, and financial transactions are some of the possible applications. By providing ubiquitous mobile broadband connectivity the standard will be strongly competitive with DSL and cable, as well as 3G cellular technologies.

Chapter 5 Concluding Remarks

We are at the beginning of a significant change in wireless systems and services as IEEE 802 wireless standards continue to evolve. Future standards will require further innovations in architectures, spectrum allocation, and protocols [B12], [B50], [B101]. Ultimately, these areas are related, and innovations in one area will lead to significant developments in other areas. Convergence among the different wireless technologies is still in its infancy. Due to technological and business issues, convergence among the different wireless standards and truly ubiquitous connectivity will require a major change. This convergence includes wireless networks seamlessly operating with other wireless networks and even with wired networks. Future standards will have highly efficient use of the wireless spectrum. High-performance physical layers are needed to provide data rates on the order of 1 Gb/s. At these data rates, the coherence bandwidths become a smaller percentage of the channel bandwidths, and there will be a greater number of resolvable paths. Clearly, OFDM is very attractive at these higher data rates. Space-time coding and smart antennas provide higher spectral efficiencies through better spatial reuse. Furthermore, advanced signal processing (such as multiuser detection) can also help attain higher spectral efficiencies. Clearly, parametric control over the modulation and coding will be available, together with power levels and other related functions. The adaptation must be quick. Both centralized and ad-hoc networking must be efficiently supported. The MAC may need to include signaling information and the protocol for this seems an attractive research topic. Note that variations of OFDM include MC-CDMA and other coded schemes. MC-CDMA applies a spreading sequence on the frequency domain. Higher-level modulations can be used in some carriers and not in others. It is possible to combine all multiple access schemes—TDMA/FDMA/CDMA is feasible. The research into the best combination of multiple access scheme, modulation, and code is expected to continue. Note that to use spectrum efficiently, in addition to the multiple-access scheme, the duplexing scheme must be considered. QoS adaptability requires the QoS to be negotiated between the stations.

Chapter 5: Concluding Remarks

In the process of communication, resources need to be assigned and managed by a multilayer adaptive radio resource management (RRM). This RRM will include not only dynamic spectrum management, but also dynamic management of the multiple access schemes. Not only TDMA and CSMA schemes will be used, but more complex time/slot/space schemes will be provided where the resource will be assigned dynamically to achieve efficient multiple access for a variety of traffic types.

A technology that allows ultimate flexibility in many respects is software radio [B109]. Software-defined radio (SDR) is a radio with enough programmability, i.e., where enough is loosely defined. The SDR Forum is the main industry group promoting SDR and developing standards in SDR technology. While current technology limitations do not allow the full potential of software radio to be realized, the availability of inexpensive SDR solutions will be one of the key enabling technologies for the future wireless systems.

Bibliography

[B1] Aad, I. and Casteluccia, C., "Differentiation mechanisms for IEEE 802.11," IEEE INFOCOM 2001.

[B2] Aboba, B., Beadles, M., "The Network Access Identifier," RFC 2486, January 1999.

[B3] Aiello, R., Brethur, V., and Kareev, U., "UWB, a solution for location awareness in TG4 applications," IEEE 802.15-03/050r0, January 2003.

[B4] Alamouti, Siavash M., "A Simple Transmit Diversity Technique for Wireless Communications," *IEEE Journal on Select Areas in Communications,* Vol. 16, No. 8, pages 1451–1458, October 1998.

[B5] ATM Forum Specification af-uni-0010.002, ATM User-Network Interface Specification, Version 3.1, September 1994.

[B6] ATM Forum Specification af-sig-0061.000, ATM User-Network Interface (UNI) Signaling Specification, Version 4.0, July 1996.

[B7] ARIB STD-T66 Second Generation Low Power Data Communication System/Wireless LAN System 1999.12.14 (H11.12.14) Version 1.0. Association of Radio Industries and Businesses (ARIB). Japan.

[B8] Baldwin, K., Webster, M., and Halford, S., "Topics in MIMO channel modeling," IEEE 802.11-03-566r0, July 2003.

[B9] Bing, Benny, *Broadband Wireless Access,* Kluwer, Norwell, MA, 2000.

[B10] Benveniste, M., "Self-configurable wireless LAN systems," IEEE 802.11-02-211r0, March 2002.

[B11] Benveniste, M., "Wireless LANs and neighborhood capture," preprint.

[B12] Berezdivin, R., Breinig R., and Topp, R., "Next-generation wireless communication concepts and technologies," *IEEE Communications Magazine,* pp. 108-116, March 2002.

Bibliography

[B13] Bluetooth Special Interest Group, "Core Specification Version 1.1," Specification of the Bluetooth System, February 22, 2001. [Bluetooth_11_Specifications_Book.pdf]

[B14] Blunk, L., Vollbrecht, J., "PPP Extensible Authentication Protocol (EAP)," RFC 2284, March 1998.

[B15] Bradner, S., "Key words for use in RFCs to Indicate Requirement Levels," RFC 2119, March, 1997.

[B16] Cali, F., Conti, M., and Gregori, E., " Dynamic tuning of the 802.11 protocol to achieve a theoretical throughput limit," IEEE/ACM Transactions on Networking, vol. 8, pp. 785-799, 2000.

[B17] Callaway, E., "TG4 battery life extension," IEEE 802.15-03/037r0, January 2003.

[B18] Cantoni, A. and Godara, L. C., "Fast Algorithm for Time Domain Broad Band Adaptive Array Processing," IEEE Trans. Aerops. Electr. Syst., AES-18, 682, September 1982.

[B19] Chayat, N., Tentative criteria for comparison of modulation methods, IEEE P802.11-97/96, Sept. 1997.

[B20] Chiasserini, C. F. and Rao, R. R., "Coexistence mechanisms for interference mitigation between 802.11 WLANs and Bluetooth," IEEE INFOCOM 2002.

[B21] Choi, W. Y. and Lee, S. K., "MAC throughput enhancement by dynamic dot11RTSThreshold," IEEE 802.11-03-0509r0, July 2003.

[B22] Clark, M., Leung, K., McNair, B., and Kostic, Z., "Outdoor IEEE 802.11 cellular networks: radio link performance," IEEE International Conference Communications, 2002.

[B23] Congdon, P., et al. "IEEE 802.1X Usage Guidelines," Internet draft, draft-congdon-radius-8021x-15.txt, July 2001.

[B24] Cooklev, T., "Dynamic bandwidth allocation and channel coding in providing QoS for wireless local area networks," International Conference Telecommunications, Papeete, Tahiti, February 2003.

[B25] Cooklev, T., Tzannes, M., Lee, D., and Lanzl, C., "Extended data rate 802.11a," IEEE 802.11-02/231r0, March 2002.

[B26] Droms, R., "Dynamic Host Configuration Protocol," RFC 2131, March 1997.

[B27] Dworkin, M., Draft recommendation for block cipher modes of operation: The CCM mode for authentication and confidentiality, NIST special publication 800-38C, Sept. 2003.

[B28] ECC "Strategic Plans for the Future. Use of the Frequency Bands 862-870 MHz and 2400-2483.5 MHz for Short Range Devices," May, 2002.

[B29] Edmonson, P. J., "Wearable computers and wireless connectivity," Wireless 2000, pp. 260-267, Calgary, Canada.

[B30] Eidson, B. and McKown, R., "WirelessMAN-SCa PHY extensions for mobility," IEEE C802.16e-03/27, May 2003.

[B31] Elders-Boll, H. and Dawid, H., "Implementation of a 480 Mb/s Viterbi decoder for 802.15.3a," IEEE 802.15-03/349r0, Sept. 2003.

[B32] ERC Recommendation 70-03, Relating to the use of Short Range Devices (SRD), April 2002.

[B33] ETSI, "Criteria for comparison," ETSI BRAN #9, July 1998.

[B34] ETSI EN 300 220-1: Electromagnetic Compatibility and Radio Spectrum Matters (ERM); Short Range Devices (SRD); Radio equipment to be used in the 25 MHz to 1000 MHz frequency range with power levels ranging up to 500 mW; Part 1: Technical characteristics and test methods.

[B35] ETSI EN 300 328-1: Electromagnetic Compatibility and Radio Spectrum Matters (ERM); Wideband Transmission systems; Data transmission equipment operating in the 2,4 GHz ISM band and using spread spectrum modulation techniques; Part 1: Technical characteristics and test conditions.

[B36]　ETSI EN 300 328-2: Electromagnetic Compatibility and Radio Spectrum Matters (ERM); Wideband Transmission systems; Data transmission equipment operating in the 2,4 GHz ISM band and using spread spectrum modulation techniques; Part 2: Harmonized EN covering essential requirements under article 3.2 of the R&TTE Directive.

[B37]　Faulhaber, G. and Farber, D. J., "Spectrum management: property rights, markets, and the commons," presented at the FCC, June 2002.

[B38]　Fahmy, N., Todd, T., and Kezys, V., "Ad hoc networks with smart antennas using 802.11-based protocols," IEEE Int. Conference Communications 2002.

[B39]　Federal Communications Commission (FCC), Document: 47 CFR, Part 15, section 15.247, USA

[B40]　Federal Communications Commission (FCC), Part 15 regulations, March 13, 2003.

[B41]　Federal Communications Commission (FCC), " Unlicensed NII devices in the 5 GHz frequency range," Federal register, vol. 62, No. 21, pp. 4649-4657, January 31, 1997.

[B42]　Federal Communications Commission (FCC), "Unlicensed NII devices in the 5 GHz frequency range," Federal register, vol. 63, No. 147, pp. 40831-40837, July 31, 1998.

[B43]　Fettweis, G. and Nitsche, G., "1/4 Gb/s WLANs," IEEE 802.11-02-320r0, May 2002.

[B44]　FIPS Pub 198, The Keyed-Hash Message Authentication Code (HMAC), Federal Information Processing Standards Publication 198, US Department of Commerce/N.I.S.T., March 6, 2002.

[B45]　Fixed wireless access system using quasi-millimeter-wave and millimeter-wave band frequencies—point-to-multipoint system, ARIB STD-T59, rev. 1, March 2000.

[B46]　Ghosh, M., Ouyang, X., Dolmans, G., "On the use of multiple antennae for 802.11," IEEE 802.11-02-180r0, March 2002.

[B47] Hartsen, J. C., "The Bluetooth radio system," *IEEE Personal Communications Magazine,* pp. 28–36, February 2000.

[B48] Heegard, Chris et. al., "High-Performance Wireless Ethernet," *IEEE Communications Magazine,* Nov. 2001, pp. 64–73.

[B49] Herrera, J., et. al., "Switched-beam antennas in millimeter-wave band broadband wireless access networks," IEEE C802.16-03/09, July 2003.

[B50] Hirano, J., Yew, T., and Aramaki, T., "Global area network concept," IEEE 802.11-02-149r0, March 2002.

[B51] Holland, G., Vaidya, N., and Bahl, P., "A rate-adaptive MAC protocol for multi-hop wireless networks," ACM SIGMOBILE, pp. 236–250, 2001.

[B52] http://www.arib.or.jp/english

[B53] http://www.bluetooth.com

[B54] http://www.etsi.org

[B55] http://www.fcc.gov

[B56] http://www.ic.gc.ca

[B57] http://www.ieee802.org

[B58] http://www.ieee802.org/11

[B59] http://www.ieee802.org/15

[B60] http://www.ieee802.org/16

[B61] http://www.wi-fi.org

[B62] http://www.x9.org

[B63] http://www.zigbee.org

[B64] Iachella, S., Deployment engineering of WLAN systems, M.S. thesis, San Francisco State University, 2003.

Bibliography

[B65] IEEE 802™-2001, IEEE Standard for Local and Metropolitan Area Networks: Overview and Architecture.[12]

[B66] IEEE 802.1Q™, 2003 Edition, IEEE Standards for Local and metropolitan area networks—Virtual Bridged Local Area Networks.

[B67] IEEE 802.1X™-2001 IEEE Standards for Local and Metropolitan Area Networks: Port-Based Network Access Control.

[B68] IEEE Std 802.11™, 1999 Edition (R2003) (ISO/IEC 8802-11: 1999), IEEE Standard for Information Technology—Telecommunications and Information Exchange between Systems—Local and Metropolitan Area Network—Specific Requirements—Part 11: Wireless LAN Medium Access Control (MAC) and Physical Layer (PHY) Specifications.

[B69] IEEE Std 802.11a™-1999 (R2003) (ISO/IEC 8802-11:1999/Amd 1:2000(E)), Information technology—Telecommunications and information exchange between systems—Local and metropolitan area networks—Specific requirements—Part 11: Wireless LAN Medium Access Control (MAC) and Physical Layer (PHY) specifications—Amendment 1: High-speed Physical Layer in the 5 GHz band.

[B70] IEEE 802.11b™-1999 (R2003) Supplement to 802.11-1999, Wireless LAN MAC and PHY specifications: Higher speed Physical Layer (PHY) extension in the 2.4 GHz band.

[B71] IEEE 802.11e™/D8, IEEE Draft Amendment to Standard for Information Technology—Telecommunications and Information Exchange Between Systems—LAN/MAN Specific Requirements—Part 11: Wireless Medium Access Control (MAC) and Physical Layer (PHY) specifications: Medium Access Control (MAC) Quality of Service (QoS) Enhancements.

[B72] IEEE 802.11F™-2003, IEEE Recommended Practice for Multi-Vendor Access Point Interoperability via an Inter-Access Point Protocol Across Distribution Systems Supporting IEEE 802.11 Operation.

---

12. IEEE publications are available from the Institute of Electrical and Electronics Engineers, Inc., 445 Hoes Lane, P.O. Box 1331, Piscataway, NJ 08855-1331 USA. IEEE publications can be ordered on-line from the IEEE Standards website: http://www.standards.ieee.org/.

[B73]   IEEE 802.11g™-2003, IEEE Standard for Information technology—Telecommunications and information exchange between systems—Local and metropolitan area networks—Specific requirements—Part 11: Wireless LAN Medium Access Control (MAC) and Physical Layer (PHY) specifications—Amendment 4: Further Higher-Speed Physical Layer Extension in the 2.4 GHz Band.

[B74]   IEEE 802.11h™-2003, IEEE Standard for Information technology—Telecommunications and Information Exchange Between Systems—LAN/MAN Specific Requirements—Part 11: Wireless LAN Medium Access Control (MAC) and Physical Layer (PHY) Specifications: Spectrum and Transmit Power Management Extensions in the 5GHz band in Europe.

[B75]   IEEE 802.11i™-2004, IEEE Standard for Telecommunications and Information Exchange Between Systems—LAN/MAN Specific Requirements—Part 11: Wireless LAN Medium Access Control (MAC) and Physical Layer (PHY) Specifications: Medium Access Control Security Enhancements.

[B76]   IEEE P802.11k™, Draft Specification for radio resource management, 2003.

[B77]   IEEE 802.12™, 1998 Edition (ISO/IEC 8802-12:1998), IEEE Standard for Information technology—Telecommunications and information exchange between systems—Local and metropolitan area network—Specific requirements— Part 12: Demand-priority access method, physical layer and repeater specifications for 100 Mb/s Operation.

[B78]   IEEE Std 802.15.1™-2002—Information Technology—Telecommunications and information exchange between systems—Local and metropolitan area networks—Specific Requirements—Part 15.1: Wireless Medium Access Control (MAC) and Physical Layer (PHY) Specifications for Wireless Personal Area Networks (WPANs).

[B79]   IEEE Std 802.15.2™-2003, IEEE Recommended Practice for Information technology—Telecommunications and information exchange between systems—Local and metropolitan area networks—Specific requirements—Part 15.2: Coexistence of Wireless Personal Area Networks with Other Wireless Devices Operating in Unlicensed Frequency Bands.

[B80]   IEEE Std 802.15.3™-2003, IEEE Standard for Information Technology—Telecommunications and information exchange between systems—Local and metropolitan area networks—Specific requirements—Part 15.3: Wireless Medium Access Control (MAC) and Physical Layer (PHY) Specifications for High Rate Wireless Personal Area Networks (WPANs).

[B81]   IEEE Std. 802.15.4™-2003, IEEE Standard for Information Technology—Telecommunications and information exchange between systems—Local and metropolitan area networks—Specific requirements—Part 15.4: Wireless Medium Access Control and Physical Layer Specifications for Wireless Personal Area Networks.

[B82]   IEEE 802.16™-2001, IEEE Standard for Local and Metropolitan Area Networks—Part 16: Air Interface for Fixed Broadband Wireless Access Systems.

[B83]   IEEE Std 802.16a™-2003, IEEE Standard for Local and metropolitan area networks—Part 16: Air Interface for Fixed Broadband Wireless Access Systems—Amendment 2: Medium Access Control Modifications and Additional Physical Layer Specifications for 2-11 GHz.

[B84]   IEEE 802.16.2™-2004, IEEE Recommended Practice for Local and metropolitan area networks—Coexistence of Fixed Broadband Wireless Access Systems.

[B85]   IETF RFC 2459: Internet X.509 Public Key Infrastructure Certificate and CRL Profile, Internet Request for Comments 2459, January 1999.

[B86]   IETF RFC 3279: Algorithms and Identifiers for the Internet X.509 Public Key Infrastructure Certificate and Certificate Revocation List (CRL) profile, Internet Request for Comments 3279, April 2002.

[B87]   IETF RFC 3280: Internet X.509 Public Key Infrastructure Certificate and Certificate Revocation List (CRL) Profile, Internet Request for Comments 3280, April 2002.

[B88]   Industry Canada, IC. Document: GL36. Ottawa, Canada.

[B89]   ISO/IEC 10038 Information technology—Telecommunications and information exchange between systems—Local area networks—Media Access Control (MAC) Bridges, (also ANSI/IEEE Std 802.1D-1993), 1993.

[B90] ISO/IEC 15802-3: 1998, Information Technology—Telecommunications and information exchange between systems—Local and metropolitan area networks—Common specifications—Part 3: Media Access Control (MAC) Bridges.

[B91] ISO/IEC 8802-3 Information technology—Telecommunications and information exchange between systems—Local and metropolitan area networks—Common specifications—Part 3: Carrier Sense Multiple Access with Collision Detection (CSMA/CD) Access Method and Physical Layer Specifications, (also ANSI/IEEE Std 802.3-1996), 1996.

[B92] ITU-R Recommendation P.530-8, Propagation data and prediction methods required for the design of terrestrial line-of-sight systems.

[B93] Jai, R., Multipath Fading Channel Modeling for High Speed PHY, IEEE P802.11-97/135, Nov. 1997.

[B94] Jeon, T., et. al., "Optimal combining of STBC and spatial multiplexing for MIMO OFDM," IEEE 802.11-03-513r0, July 2003.

[B95] Johnson, D. and Dudgeon, D., *Array Signal Processing,* Prentice-Hall, Englewood Cliffs, NJ. 1993

[B96] Jun Li, Weinstein, S., Zhang, J., and Tu, N., "Public access mobility LAN: extending the wireless Internet into the LAN environment," *IEEE Trans. Wireless Communications,* Jun. 2002, pp. 22–30.

[B97] Kaitz, T., et. al., "Initial PHY proposal for 802.16e," IEEE C802.16e-03/03, January 2003.

[B98] Kamerman, A. and Monteban, L., "WaveLAN-II: A high-performance wireless LAN for the unlicensed band," *Bell Labs Technical Journal,* pp. 118–133, 1997.

[B99] Kitroser, I., "802.16e handoff description," IEEE C802.16e-03/55, Sept. 2003.

[B100] Kleindl, G., "Signaling for adaptive modulation," IEEE 802.11-03-283r0, May 2003.

[B101] Khun-Jush, J., "Integration of WLAN and wide area mobile networks," IEEE 802.11-02-106, January 2002.

[B102] Kowalski, J., "Link measurement results give thumbs-up to MAC FEC!," IEEE 802.11-01/422r0, July 2001.

[B103] Lewis, B. "Mesh networks for fixed broadband wireless access," IEEE C802.16-03/10r1, July 2003.

[B104] Liberti, J.C. and Rappaport, T.S., *Smart Antenna for Wireless Communications,* Prentice-Hall, Englewood Cliffs NJ, 1999.

[B105] Lien, J., Wu, C., Liu, H., and Lung, Y., "An adaptive multirate IEEE 802.11 wireless LAN," International Conf. On Inf. Networking, pp. 411–418, 2001.

[B106] McCann, S., "Proposed wireless interworking group (WIG) baseline document," IEEE 802.11-03-004r0, January 2003.

[B107] McCorkle, J., "DS-CDMA—The technology of choice for UWB," IEEE 802.15-03/277r0SG3a, July 2003.

[B108] Milewski, A., "Periodic Sequences with Optimal Properties for Channel Estimation and Fast StartUp Equalization," IBM J. Res. Develop., Vol. 37, No.5, Sept. 1983, pp. 426–431.

[B109] Mitola, J., *Software radio architecture,* John Wiley, New York, NY, 2002.

[B110] Mulhens, O., "QoS-enabled multihop meets high data rate," IEEE 802.11-02-333r0, May 2002.

[B111] Mulhens, O., "Cluster-based multihop networking with WPANs," IEEE 802.15-02/312r0, July 2002.

[B112] Nasipuri, A., et al, "A MAC protocol for mobile ad-hoc networks using directional antennas," Proc. IEEE Wireless Communications and Networking Conf (WCNC), Sept. 2000.

[B113] NIST FIPS Pub 197: Advanced Encryption Standard (AES), Federal Information Processing Standards Publication 197, US Department of Commerce/N.I.S.T., November 2001.

[B114] O'Hara, Robert and Petrick, Al, *IEEE 802.11 Handbook: A Designer's Companion,* IEEE Standards Information Network/IEE Press, 1999.

[B115] Okamoto, G. and Xu, G., "Throughput multiplication of wireless LANs for multimedia services: spread-spectrum with SDMA," IEEE Vehicular Technology Conference, April 1996.

[B116] Pahlavan, A. Levesque, *Wireless Information Networks,* John Wiley, New York, NY, 1995.

[B117] Pazhyannur, R., "WLAN-cellular interworking," IEEE 802.11-03-193r1, March 2003.

[B118] PKCS#1 v2.1, RSA Cryptography Standard, RSA Laboratories, June 14, 2002.9

[B119] Porcino, D., Hirt, W., "Ultra-wideband radio technology: potential and challenges ahead," *IEEE Communications Magazine,* vol. 41, pp. 66–74, July 2003.

[B120] Prabhu, B. J. and Chockalingam, A., "A routing protocol and energy efficient techniques in Bluetooth scatternets," IEEE Int. Conference Communications, 2002.

[B121] Rappaport, T., *Wireless Communications: Principles and Practice, Prentice-Hall,* Upper Saddle River, NJ, 2002.

[B122] Recommendation ITU-R P.1238-1, Propagation data and prediction methods for the planning of indoor radiocommunication systems and radio local area networks in the frequency range 900 MHz to 100 GHz. Question ITU-R 211/3.

[B123] Regnier, J., "Benefits of smart antennas in 802.11 networks," IEEE 802.11-03-025r0, 2003.

[B124] Rigney, C., "RADIUS Accounting," RFC 2866, June 2000.

[B125] Rigney, C., Rubens, A., Simpson, W., Willens, S., "Remote Authentication Dial In User Service (RADIUS)," RFC 2865, June 2000.

[B126] Rigney, C., Willats, W., Calhoun, P., "RADIUS Extensions," RFC 2869, June 2000.

[B127] Rivest, R., Dusse, S., "The MD5 Message-Digest Algorithm," RFC 1321, April 1992.

[B128] Standard specification for telecommunications and information exchange between roadside and vehicle systems—5 GHz band Dedicated Short-Range Communications (DSRC) Medium-Access Control (MAC) and Physical Layer (PHY) specifications, American Society for Testing and Materials, ASTM E2213, Nov. 2002, Aug. 2003.

[B129] SEC 1: Standards for Efficient Cryptography, Elliptic Curve Cryptography, Version 1.0, Certicom Research, September 20, 2000.

[B130] Siep, Tom, *An IEEE Guide: How to Find What You Need in the Bluetooth Spec.,* IEEE Standards Information Network/IEE Press, 2000

[B131] Seggal, Y., et. al., "OFDMA modification for mobility," IEEE C802.16e-03/25r2, May 2003.

[B132] Stallings, W., *Cryptography and Network Security: Principles and Practice,* Prentice-Hall, Upper Saddle River, NJ, 2002.

[B133] Stanwood, K. and Marks, R., "A proposal to add point-to-point option to 802.16 MAC," IEEE 802.16-03/11, July 2003.

[B134] Unbehaun, M., "Coverage Planning for Indoor Wirelss LAN Systems," ICC 2001, Helsinki, Finland.

[B135] Verdu, S., *Multiuser Detection,* Cambridge University Press, 1998.

[B136] Viterbi, A. J., Wolf, J. K., Zehavi, E., and Padovani, R., "A pragmatic approach to trellis-coded modulation," *IEEE Communications Magazine,* July 1989, pp. 11–19.

[B137] Veeraraghavan, M., Cocker, N., and Moors, T., "Support of voice services in IEEE 802.11 wireless LANs," IEEE INFOCOM 2001.

[B138] Veeraraghavan, M., Cocker, N., Moors, T., "Support of voice services in IEEE 802.11 wireless LANs," IEEE INFOCOM 2001.

[B139] Welborne, M., et. al. "Merger #2 proposal DS-CDMA," IEEE 802.15-03/334r3, Sept. 2003.

[B140] Whiting, D., Housely, R., Ferguson, N., "Counter with CBC-MAC (CCM)," RFC 3610, Sept. 2003.

[B141] Win, M. Z., Cassioli, D., and Molisch, A., "The ultra-wide bandwidth indoor channel: from statistical modeling to simulations," IEEE 802.15-02/284r0SG3a, June 2002.

[B142] Win, M. Z., Cassioli, D., and Molisch, A., "A statistical model for the UWB inddor channel," IEEE 802.15-02/285r0SG3a, June 2002.

[B143] Wolf, J. K. and Zehavi, E., "P2 codes: Pragmatic trellis codes utilizing punctured convolutional codes," *IEEE Communications Magazine,* February 1995, pp. 94–99.

[B144] Wu, H., et. al. "IEEE 802.11 distributed coordination function (DCF) analysis and enhancement," IEEE Int. Conf. Communications, 2002.

[B145] Xao, Y. and Rosdahl, J., "Throughput analysis for 802.11a higher data rates," IEEE 802.11-02-138r0, March 2002.

[B146] Yanover, V., et. al., "Some thoughts for the New Concepts forum," IEEE C802.16-03/13, July 2003.

[B147] Yeh, J.-Y. and Chenm, C., "Support of multimedia services with the IEEE 802.11 MAC protocol," IEEE ICC 2002.

[B148] Yergeau, F., "UTF-8, a transformation format of Unicode and ISO 10646," RFC 2044, October 1996.

[B149] Yu, H., et. al., "MIMO OFDM with antenna selection," IEEE 802.11-03-514r0, July 2003.

[B150] Yu, H., "HDR 802.11a solution using MIMO-OFDM," IEEE 802.11-02-294 r0, May 2002.

[B151] Zorn, G., Leifer, D., Rubens, A., Shriver, J., Holdrege, M., Goyret, I., "RADIUS Attributes for Tunnel Protocol Support," RFC 2868, June 2000.

[B152] Zorn, G., Mitton, D., Aboba, B., "RADIUS Accounting Modifications for Tunnel Protocol Support," RFC 2867, June 2000.

Glossary

access point: A special station that allows a network to connect with another network.

acknowledgment field: Bit present in every frame that will, if set, require acknowledgment.

active service flow: A service flow requesting and being granted bandwidth for transporting data packets.

adaptive modulation: Different modulation for each subscriber station.

ad-hoc traffic indication message (ATIM) window: The amount of time after every beacon that a device must stay awake.

admitted service flow: A service flow with resources assigned but not yet activated.

alternating wireless medium access: Collaborative coexistence method where two networks with separate links have their transmissions scheduled in different time segments.

authentication key: Secret key derived in initialization and shared between two or more parties as the base for all their security transactions. The four kinds are unit, combination, temporary, and initialization. Also called link key.

authentication node: In mesh mode, a node that verifies the digital certificate of new stations, acting as a BS.

authorization module: BS module acting as QoS manager, approving or denying changes to the QoS parameters.

bandwidth stealing: When a station uses bandwidth already given for another request.

beacon frame: Management frame that maintains the synchronization of the local timers in the stations and delivers protocol-related parameters.

Glossary

bridges: Layer 2 devices used to connect network segments at the MAC layer.

broadband: In ITU terminology, transmission rates greater than 1.5 Mb/s.

bursty traffic: Signal traffic with messages of arbitrary length separated by intervals of random duration, such as office data communication (e.g., e-mail, Internet access, file sharing).

CAT5: Special cables required for Ethernet wiring.

Channel Estimation Interval (CEI): Vendor-specific interval, typically 20 ms, that defines how often the BS must ask an SS for updated channel state information.

channel time assignment (CTA): Dedicated time slot from the parent piconet that determines the piconet identifier of child or neighbor piconets.

ciphertext: In a cryptographic system, an encrypted message.

contention window (CW): The maximum extent of the backoff time.

contention-free period: Time period for WLANs in which a point coordinator is active.

data region: OFDMA slot with time and subchannel dimensions. The time dimension denotes consecutive OFDM symbols in the time domain, and the subchannel dimension denotes blocks within the OFDM symbols.

delay spread: Difference between the arrival times of the first and last paths of signals.

digital modulation: As defined by the FCC, when the combination of data rate, coding, and modulation method has a 6 dB bandwidth greater than 500 kHz and a maximum transmitted spectral density of less than +8 dBm/3 KHz.

distribution services: Services that manage traffic when an access point is present: association, disassociation, distribution, integration, and reassociation.

donut pattern: Omnidirectional pattern radiating equally in all directions around an axis or azimuth in a horizontal plane with a circular pattern outward.

Doppler spread: Measure of spectral broadening caused by relative mobility.

Glossary

encapsulation protocol: One of two protocols of MAC security sublayer. Defines a set of supported cryptographic suites and the rules for applying them to a MAC payload.

encapsulation: The way cryptographic data is constructed from plaintext data.

Encryption Control: Bit field in the MAC header indicating whether the payload is encrypted.

fading: Fluctuation in the received signal, the result of the randomness with which the multipath components add at the receiver. Fading depends on both the spatial position of the receiver, and on the frequency of transmission.

FCH OFDM symbol: Symbol following the long preamble on downlink that conveys the burst parameters and other control information.

Fresnel zone: Area around the optical line of sight that must be free of obstacles to avoid random signal loss.

guaranteed time slots (GTSs): Dedicated slots allocated by the network-coordinator that comprise the contention-free period. Used for periodic traffic.

HiperAccess: ETSI technology for wireless broadband access, much like IEEE 802.16.

HiperLAN: ETSI technology for wireless LAN, much like IEEE 802.11.

interframe space (IFS): Time interval between frames. The five types are short IFS (SIFS), point coordination function IFS (PIFS), distributed IFS (DIFS), arbitration IFS (AIFS), and extended IFS (EIFS).

interleaving: A technique to reduce the statistical dependence of errors in which the symbols in one code block are not transmitted in consecutive order, but instead are interspersed among other symbols.

interworking: The ability to perform a task by two systems using different set of rules.

jitter: Delay variation.

link key: See *authentication key*.

link-level authentication: Open-system and shared-key authentication, defined in 802.11 in 1997.

log-normal shadowing: An effect showing random path loss according to a Gaussian (or normal) distribution for the free-space path loss model, which provides the average value.

multipath: Phenomenon by which a signal will reach the receiver by traveling multiple paths.

multipoint: Architecture of fixed BWA system that includes base stations, subscriber stations and, in some cases, repeaters. May be either point-to-multipoint (PMP) and multipoint-to-multipoint (MP-MP).

neighbor: (1) Stations with which a node has direct links, "one hop" away from node. Architectural element of mesh systems. (2) Dependent piconet whose coordinator does not participate in the parent piconet.

nonce: Abbreviation for "number once," the random number sent by a device in an authentication effort.

offline classification: Channel classification when a piconet is not established or is on hold. May involve RSSI measurements.

packet traffic arbitration: Collaborative coexistence method where 802.11b and 802.15.1 networks can decide what type of packet to send to minimize the interference between them.

periodic traffic: Signal traffic with an approximately constant data rate and a small interarrival variance between messages, such as voice and video.

per-packet key: A concatenation of an initialization vector (IV) and a private encryption key. Also called a seed.

piconet: Basic ad-hoc network of a wireless personal area network.

plaintext: In a cryptographic system, a message to be encrypted.

poll interval: Time interval equal to the maximum time between subsequent transmissions from the master to a particular slave on an ACL link, used to provide dynamic bandwidth allocation and latency.

Glossary

power control: Adjustment of power levels in a monotonic sequence, required for class 1 equipment.

privacy key management (PKM): One of two protocols of the security sublayer. Describes how the BS distributes and synchronizes keys in a secure fashion.

provisioned service flow: Service flow with an assigned Service Flow ID but no reserved resources. A flow not immediately activated. Also called a deferred service flow.

puncturing: A method to reduce the number of codeword bits and increase the rate of the code.

receiver sensitivity: Certain minimum value of the power of a received signal. Above this point, communication is possible.

repeaters: Layer 2 devices used to connect network segments at the PHY layer.

routers: Devices used to connect network segments at the network layer.

Rx/Tx transition gap: Interval between the uplink burst and the subsequent downlink burst that allows time for the BS and SS to switch modes (BS from receive to transmit, SS from transmit to receive). Begins on physical slot boundary.

scatternet: A collection of overlapping piconets.

security association: Set of security information shared by a BS and SS. Includes traffic encryption keys and cipher block chaining (CBC) initialization vectors. May be primary, static, or dynamic.

seed: See *per-packet key*.

service access point: The interface between any two protocol layers.

service flow: Unidirectional flow of packets provided to a particular QoS according to the QoS parameter set.

session: The finite time interval in which units participate in a piconet.

slot time: The time unit of the backoff timer.

Glossary

sniff mode: Mode used to reduce the duty cycle of the slave, where the master can transmit in regularly spaced, specified time slots.

snooze mode: Low-power mode to be used when there are no pending messages and no activity from the devices. The coordinator stops transmitting but keeps track of superframe boundaries and listens for messages.

station services: Services that are part of every station: authentication, deauthentication, key distribution, data authentication, replay protection, privacy, and delivery of data.

station: A logical device that participates in a network. Consists of a physical layer and a medium access control layer.

superframe: Frame with two periods: contention-free (when DCF is active) and contention (when PCF is active).

targeted beacon transmission time (TBTT): The intended arrival point of the next beacon frame, announced in every beacon frame.

transition security network: Inherently insecure network in which some equipment supports advanced security algorithms and some does not.

transmission opportunity: The interval of time when a particular station has the right to initiate transmissions onto the wireless medium.

turnaround time: The time for a device to switch from receive to transmit mode (RxTx turnaround time), or from transmit to receive mode (TxRx turnaround time).

Tx/Rx transition gap: Interval between the downlink burst and the subsequent uplink burst that allows time for the BS and SS to switch modes (BS from transmit to receive, SS from receive to transmit). Begins on physical slot boundary.

U-NII bands: Unlicensed national information infrastructure bands, 5.15–5.825 GHz.

unscheduled service periods: Service periods in response to a poll.

WirelessHUMAN: Protocol for operating in the unlicensed frequencies in the 2–11 GHz bands.

Index

Numerics

802.11 3, 19, 20, 22, 24, 45, 107, 110, 133, 134, 135, 144, 165, 166, 167, 168, 169, 170, 171, 176, 177, 178, 179, 183, 188, 195, 199, 207, 214, 216, 217, 228, 247, 271, 290, 304, 311, 316
  MAC 70–96
  Physical layer 46, 96
  Security 49–66
802.11a 107, 110
802.15 19, 21, 24, 70, 135–173, 176, 177, 178, 179, 180, 181, 182, 183, 187, 188, 192, 195, 199, 201, 202, 204, 207, 214, 215, 216, 217, 229, 271, 316
  MAC layer 181–195
  Physical layer 137–138, 195–201
802.16 19, 24, 70, 228–331
802.20 331, 332

A

Access point 48, 52, 55, 57, 66, 67, 68, 70, 77, 78, 79, 81, 82, 112, 117, 122, 233
ACL 145, 146, 147, 148, 150, 151, 153, 163, 168, 169, 171, 172, 174, 186, 213
Advanced antenna systems 32, 261
Advanced encryption standard 61, 214, 215, 216
Antenna 120–123, 261–265, 302, 323, 326
Antenna diversity 302
Antenna polarization 120, 323
Attenuation 29

Authentication 50, 52, 54, 55, 57, 61, 70, 158, 186, 248, 250, 251, 252, 253, 261
Automatic repeat request 40, 142, 149, 171, 236, 257, 265, 266, 267, 289, 296

B

Backoff 75, 208, 247
Bandwidth 235, 247, 326
Beacon 91, 94, 213
BER 71, 169, 170, 171, 172, 176, 177, 270, 311
Binary phase-shift keying (BPSK) 109, 110
Bit 71
Bluetooth 18, 24, 135–167, 171–179, 201
  Connection modes 151–152
  Physical layer 137–138, 195–201
  Security 154–162
BPSK 101, 106, 109, 110, 195, 207, 291, 296, 307, 315
BS, base station 226, 227, 237, 238, 239, 240, 241, 243, 244, 245, 246, 247, 248, 249, 250, 251, 252, 253, 254, 255, 256, 257, 258, 259, 260, 261, 262, 263, 264, 268, 269, 271, 273, 277, 278, 283, 284, 285, 286, 287, 309, 310, 313, 314, 317, 318, 319, 320, 326, 327, 328
BWA 4, 225, 226, 227, 287, 316, 317, 318, 319, 323, 324, 325, 327, 328

C

CCA 110

Index

CCK 99
CINR, carier-to-interference-plus-noise-ratio 302, 311
Coexistence 117, 164–178, 216, 316, 324
Collision 20
Comparing WPAN and WLAN 133–135
Connection 134, 145, 237, 241, 249, 251
Contention window 73, 86, 87
Contention-free 77, 79, 80, 84, 91, 187, 188, 189, 191, 192, 193, 212
Convolutional coding 300, 305
CSMA 20, 37, 46, 70, 73, 78, 112, 117, 134, 144, 187, 188, 202, 207, 208, 209, 212, 246, 336
CTS 76, 121

D

DFS 47, 111, 257, 267, 268, 287, 322, 323
Downlink 257, 263, 273, 274, 275, 278
DSSS 8, 15, 97, 124, 207

E

EIRP (effective radiated power relative to isotropic) 25, 177, 285, 323, 325
Encryption 50, 60, 66, 159, 249, 251, 255, 256
Enhanced distributed coordination function 85, 89, 90, 91
Error 71, 74, 177
ETSI 14, 15, 24

F

Fading 29
Fast Fourier transform (FFT) 107, 109, 110
FCC 5, 6, 8, 10, 11, 14, 15, 45, 47, 201
FDD 234, 235, 264, 271, 272, 273, 274, 287, 288, 307, 308, 313, 318, 329
FDMA 335

FEC, forward error control 142, 143, 144, 147, 148, 161, 171, 172, 272, 274, 275, 276, 277, 281, 284, 289, 296, 304, 305, 311, 313, 315
FFT 106, 107, 109, 110, 304, 312, 313
Frequency band 116, 195, 206
Frequency hopping 9, 15, 41, 173

G

Gray coding 282, 291, 296, 298, 299, 305

H

Hybrid coordination function 84, 89, 90, 91, 93, 94

I

IETF 52, 54, 68
Interference 165, 316, 317, 320, 325
Internet 1, 37, 48, 52, 54, 133, 178, 225, 232, 233, 270, 330, 331
ISM 7, 8, 10, 13, 14, 20, 21, 103, 116, 137, 165, 166, 195, 202, 204, 205
ITU 24, 149, 225, 271, 326

M

MAC 19, 20, 21, 36, 37, 39, 40, 41, 42, 43, 46, 47, 49, 52, 55, 61, 66, 83, 99, 110, 136, 137, 144, 165, 167, 176, 178, 179, 180, 199, 209, 215, 233, 272, 274, 276, 277, 278, 285, 302, 305, 335
  802.11 70–96
  Fixed broadband wireless (802.16) 228, 231–268, 269
  High-rate WPAN (802.15.3) 181–195
  Low-rate WPAN (802.15.4) 203, 207–216
  Security 213–216, 240, 248–256
Medium-access control (MAC) 110

Index

MIC 60, 61, 62, 214
Multimedia 228

O

OFDM 9, 32, 41, 104, 106, 107, 109, 110, 113, 195, 229, 287, 303, 304, 307, 308, 309, 311, 312, 313, 314, 315, 335
Orthogonal frequency-division multiplexing (OFDM) 107, 109, 110

P

Packet 37, 148, 178, 232
Packet binary convolutional coding 47, 99, 101, 113
PER 166, 170, 172
Periodic traffic 37
Physical layer 19, 21, 40, 41, 42, 46, 72, 86, 99, 104, 136, 137, 167, 176, 178, 179, 182, 184, 188, 193, 194, 195, 196, 199, 200, 204, 206, 207, 228, 230, 238, 247, 256, 260, 262, 263, 269, 270, 271, 272, 285, 286
   802.11 96
   802.15.3 195–201
   Bluetooth (802.15.1) 137–138
   Fixed broadband wireless (802.16) 269–316

Q

QAM 106, 109, 110, 138, 195, 196, 197, 198, 272, 274, 281, 282, 284, 291, 292, 293, 294, 295, 296, 297, 298, 299, 311, 327
QPSK 100, 101, 106, 195, 196, 199, 200, 205, 272, 274, 275, 281, 284, 291, 296, 301, 302, 305, 308, 309, 311, 327
Quadrature amplitude modulation (QAM) 109, 110

Quality of Service 46, 47, 57, 71, 80, 83, 84, 89–94, 96, 135, 141, 144, 151, 174, 179, 181, 187, 190, 194, 202, 212, 218, 229, 233, 236, 237, 241, 244, 245, 247, 335

R

Radio frequency 7, 25, 72, 122, 141, 149, 167, 204, 205, 239, 281, 282, 286, 287, 316
Randomization, scrambling 276, 284
Receiver 311, 325
Reed-Solomon 277, 278, 281, 289, 290, 305
RTS 76, 92, 93, 121, 193

S

SCO 145, 146, 147, 149, 150, 151, 153, 163, 168, 169, 171, 172, 174
SNR, signal-to-noise ratio 166, 229, 261, 268, 281, 304
Software radio 336
Spectrum 5, 6
Speech 2, 84, 149, 245
Spread-spectrum 8
SS, subscriber station 226, 227, 234, 237, 238, 239, 240, 241, 243, 244, 245, 246, 247, 248, 249, 250, 251, 252, 253, 254, 255, 257, 258, 260, 262, 263, 264, 268, 269, 271, 273, 274, 277, 278, 281, 283, 284, 285, 291, 309, 311, 313, 314, 317, 318, 319, 320, 326, 327, 328
Standards 13, 17, 18, 19, 20, 21, 133

T

TCM, trellis coded modulation 199, 289, 290, 291, 292, 293, 294, 295
TCP/IP 1, 244

TDD 144, 154, 235, 236, 260, 264, 271,
    273, 274, 287, 288, 307, 308, 313,
    329
TDMA 37, 38, 187, 189, 230, 271, 272,
    274, 275, 289, 301, 335, 336
TKIP 54, 61
TPC 47, 111, 257, 299, 322, 323
Turbo coding 280, 300

U
UDP 68, 69
UNII 47, 104, 112
Unlicensed 8, 13
Uplink 244, 257, 263, 283, 285

V
Video 84

W
WEP 58, 59
WLAN 4, 19, 20, 45, 46, 58, 67, 72, 112,
    113, 117, 122, 124, 133, 134, 165,
    167, 168, 171, 227, 323, 331
WPAN 4, 19, 21, 133, 227, 331
    Coexistence 164–178
    High Rate 179–201
    Low Rate 201–218